Nanotechnology Science and Technology

Bioencapsulation in Silica-Based Nanoporous Sol-Gel Glasses

NANOTECHNOLOGY SCIENCE AND TECHNOLOGY

Additional books in this series can be found on Nova's website at:

https://www.novapublishers.com/catalog/index.php?cPath=23_29&seriesp=Nanotechnology+Science+and+Technology

Additional e-books in this series can be found on Nova's website at:

https://www.novapublishers.com/catalog/index.php?cPath=23_29&seriespe=Nanotechnology+Science+and+Technology

NANOTECHNOLOGY SCIENCE AND TECHNOLOGY

BIOENCAPSULATION IN SILICA-BASED NANOPOROUS SOL-GEL GLASSES

BOUZID MENAA
FARID MENAA
AND
OLGA SHARTS

Nova Science Publishers, Inc.
New York

Copyright © 2010 by Nova Science Publishers, Inc.

All rights reserved. No part of this book may be reproduced, stored in a retrieval system or transmitted in any form or by any means: electronic, electrostatic, magnetic, tape, mechanical photocopying, recording or otherwise without the written permission of the Publisher.

For permission to use material from this book please contact us:
Telephone 631-231-7269; Fax 631-231-8175
Web Site: http://www.novapublishers.com

NOTICE TO THE READER

The Publisher has taken reasonable care in the preparation of this book, but makes no expressed or implied warranty of any kind and assumes no responsibility for any errors or omissions. No liability is assumed for incidental or consequential damages in connection with or arising out of information contained in this book. The Publisher shall not be liable for any special, consequential, or exemplary damages resulting, in whole or in part, from the readers' use of, or reliance upon, this material.

Independent verification should be sought for any data, advice or recommendations contained in this book. In addition, no responsibility is assumed by the publisher for any injury and/or damage to persons or property arising from any methods, products, instructions, ideas or otherwise contained in this publication.

This publication is designed to provide accurate and authoritative information with regard to the subject matter covered herein. It is sold with the clear understanding that the Publisher is not engaged in rendering legal or any other professional services. If legal or any other expert assistance is required, the services of a competent person should be sought. FROM A DECLARATION OF PARTICIPANTS JOINTLY ADOPTED BY A COMMITTEE OF THE AMERICAN BAR ASSOCIATION AND A COMMITTEE OF PUBLISHERS.

LIBRARY OF CONGRESS CATALOGING-IN-PUBLICATION DATA

Bioencapsulation in silica-based nanoporous sol-gel glasses / Bouzid Menaa [et al.].
 p. ; cm.
Includes bibliographical references and index.
ISBN 978-1-60876-989-6 (softcover)
1 Protein folding. 2. Biocolloids. 3. Silica gel. I. Menaa, Bouzid.
[DNLM: 1. Protein Folding. 2. Biocompatible Materials--chemistry. 3.
Glass. 4. Nanocomposites. 5. Organosilicon Compounds--chemistry. 6.
Silicon Compounds--chemistry. QU 55.9 B615 2009]
QP551.B4693 2009
612'.01575--dc22

2009052729

Published by Nova Science Publishers, Inc, ✢ *New York*

Contents

Preface		vii
Abstract		ix
Chapter 1	Introduction	1
Chapter 2	Bioencapsulation via Sol-Gel Process in Silica-Based Materials: Method, Materials, Bioapplications, and Characterization Techniques	5
Chapter 3	Parameters Influencing the Protein Conformation in Nanoporous Silica-Based Sol-Gel Glasses	19
Chapter 4	Enhancing the Protein Folding by Introducing and Associating Hydrophobic and Steric Effects in Modified Silica-Based Porous Glasses	39
Chapter 5	Emerging Techniques for a Better Understanding of Protein Interactions and Conformations in Nanoporous Sol-Gel Glasses	43
Conclusion		47
References		51
Index		67

PREFACE

In this book, the authors report the recent results on the influence of different parameters (such as surface hydration, hydrophobicity, solute effects, thermal stability, porosity and macromolecular crowding) on the protein conformation based on the design and the characterization of nanoporous silica-based materials containing different functional groups (e.g., hydrophobic alkyl, phosphate and fluorinated groups). The enhancement of the protein folding, owing the physical properties and microstructure of the host matrix induced by the nature of the functional groups and the siloxane network, play a major role on the protein biological activity and, therefore, on the development efficient bionanodevices such as biocatalysts, sensors, drug delivery systems or implanted devices. This book enlightens the bioapplications of silica-based glasses prepared from functionalized organosilane precursors used to mimic and to understand the different factors that influence the protein-folding process.

ABSTRACT

Organic-inorganic nanoporous silica sol-gel glasses represents the ideal support for protein bio-encapsulation and for study of the different factors influencing the protein-folding process in a crowded environment. Due to the facile silica surface, modifications with desired Si-substituted organic groups from organosilanes precursors, organically modified "wet-aged" silica-based glasses obtained via the sol-gel process, can be used as host materials to mimic the crowded environment of the proteins and cells that can be found in the cytoplasm, for instance. Numerous studies to date showed that silica-based nanoporous glasses can stabilize bioactive proteins. However, it is important to know about the different factors affecting the protein stability and, therefore, its properly folded state. In this chapter, we report the recent results on the influence of different parameters (such as surface hydration, hydrophobicity, solute effects, thermal stability, porosity, macromolecular crowding) on the protein conformation based on the design and the characterization of nanoporous silica-based materials containing different functional groups (e.g., hydrophobic alkyl, phosphate and fluorinated groups). The enhancement of the protein folding owing the physical properties and microstructure of the host matrix induced by the nature of the functional groups and the siloxane network play a major role on the protein biological activity and, therefore, to the development efficient bionanodevices such as biocatalysts, sensors, drug-delivery systems or implanted devices. This chapter enlightens the bioapplications of silica-based glasses prepared from functionalized organosilane precursors used to mimic and to understand the different factors that influence the protein-folding process.

Keywords: organosilane precursors, nanoporous sol-gel glasses; silica-based biomaterials protein-folding, circular dichroism spectroscopy, surface hydration; crowding effects.

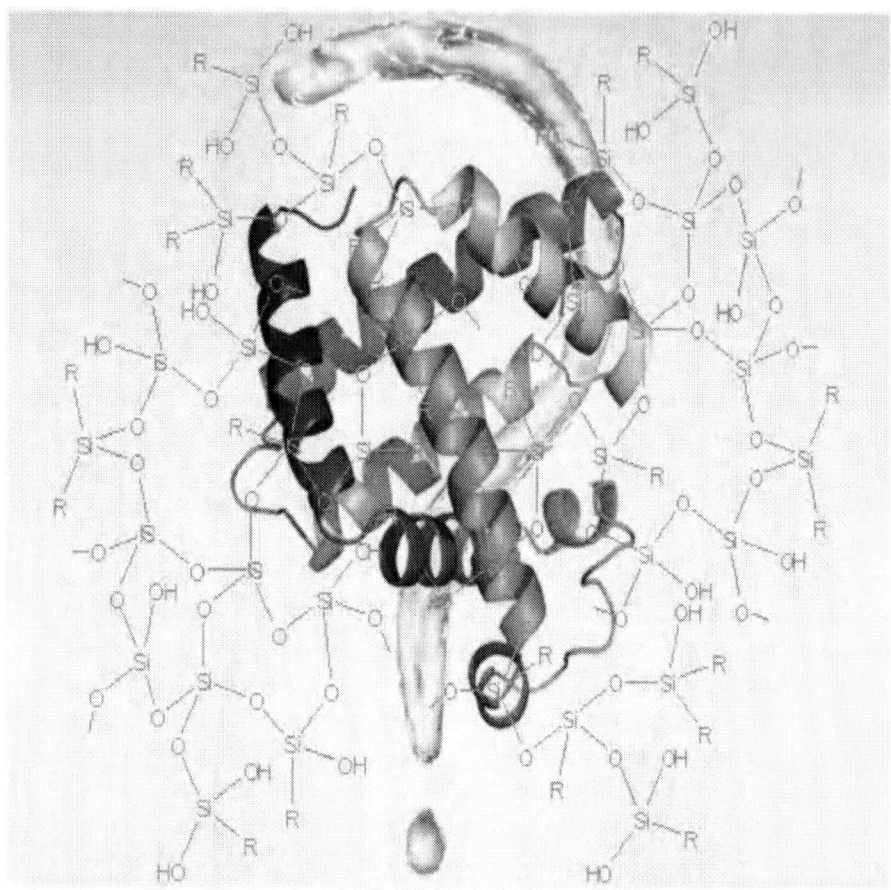

Figure illustrating the chapter content with the fundamental question that we are tempting to solve experimentally: what is the importance of silica surface modification nanoporous silica-based sol-gel glasses prepared from functionalized organosilane precursors on the parameters (such as crowding environment, surface hydration, hydrophobic effect) affecting the conformation of encapsulated protein? This chapter resumes the part of the work and results published by the authors [1] that aimed to spread a broader knowledge in the scientific community interested in the development of bioactive and biocompatible materials as well as biochemists who are eager to understand the protein-folding.-unfolding processes.

Chapter 1

1. INTRODUCTION

Inorganic and organic-inorganic hybrid porous silica-based materials constitute the ideal support to study the protein folding as a function of its microlocal environment. Due to the facile silica surface modifications, and particularly with Si-substituted organic groups, organically modified "wet-aged" based silica-glasses can be used as host materials to mimic the crowded environment [1-3] of the proteins and cells that can be found in the cytoplasm, for instance [4-6]. The study on protein conformation in biocompatible materials is very important in materials sciences for the development of new and efficient silica-based biomaterials (e.g., sensors, drug delivery systems or implanted devices) [7-12] but also in medicine, as it is well known that misfolded/unfolded proteins and disturbed protein-protein interactions are the cause of devastating diseases (e.g., Alzheimer's, Huntington's, Diabetes, etc.) [13]. This chapter results on a continuous effort to understand the protein-folding process and to find out what are the important factors related to the protein environment and how they affect its conformation in order to control them.

The host materials are generally synthesized via hydrolysis and poly condensation from alkoxysilanes precursors via the sol-gel method [14]. The encapsulation of the proteins benefits from the mild conditions of the sol-gel method as the process can be performed at room temperature to avoid the denaturation and thermal decomposition of the protein. Moreover, optically transparent biomaterials can facilitate the structure determination of the protein in the glass via spectroscopic techniques such as fluorescence or circular dichroism spectroscopy (CD) described in this chapter. Several solid-state characterization techniques briefly described in this chapter permit also to enlighten the structure and properties of the host matrix, owing to its surface modification with functional organic groups. Indeed, the protein conformation can be associated with its surrounding environment that is tuned by the design of the organically modified nanoporous sol-gel glass. Indeed, the determination of the protein conformations as function of the microlocal environment of the protein is fundamental for both the improvement and development of functional bioactive materials in which the stability and properly folded state of the protein are required for its biochemical activity and also to understand the protein-folding process by tuning the physical properties of the host matrix (macromolecular crowding, porosity, surface hydration effects, hydrophobicity etc.) affecting the protein folding and stability [5,6,15-18].

The protein encapsulation in porous sol–gel glasses [19-22] has been widely applied to heterogeneous biocatalysis applications [23-30] and to the development of solid state optical, electrochemical biosensors [7-12]. A number of reports indicate that sol-gel glass encapsulation retains the activity of a wide variety of enzymes. Most of the studies to date concerning bio-encapsulation used tetramethoxysilane-based (TMOS) glass or the organically modified TMOS glass system with $RSi(OCH_3)_3$ (R = functional organic group), which is the monoalkyltrimethoxysilane characterizing the organosilane in [TMOS : $RSi(OCH_3)_3$] systems. Many reported the catalytic activity of the proteins encapsulated that associate with lipid interfaces or proteins that bind nonpolar ligands, such as lipases [23-29,31]. It is known that lipases prefer a hydrophobic environment, and the studies on the enzyme activity showed their efficiency using silica-based sol-gel glasses using functionalized organosilane precursors containing hydrophobic organic groups.

Apomyoglobin (apoMb) is also a model protein that has been used for decades for the study of protein-folding/unfolding process in solution [32-34]. The holoprotein is very stable, but once its heme has been removed to form apomyoglobin, it is easily unfolded, and several partially folded states can be populated depending on its surrounding environment. The protein is then ideal to probe the change of the protein conformation as function of the different properties of the host matrix. In our recent studies [2,3,35], the protein was encapsulated in different glass systems [(100-x) TMOS: (x) $R_nSi(OCH_3)_{4-n}$, n = 1, 2, 3, R = alkyl, vinyl, fluorine etc.] obtained for different molar composition and including also the unmodified TMOS glass (x = 0) as control. They serve also as ideal host matrices to probe the microlocal environment change influencing the protein conformation. These glass systems can illustrate the influence of -Si-O-Si- silicon linkage in which some of the bridging oxygens of a SiO_4 tetrahedron are substituted by one-, two- or three-functional methyl groups, respectively [2,35]. We can tune the properties of the host matrix by attaching different functional organic groups at the silicon using organosilane modifiers. This modification of the host matrix surface leads to a change on surface hydration and water thermodynamics, hydrophobicity, hydrogen bondings, crowding environment and porosity. This will give us some insight on the determinant factors influencing the protein conformation in these nanoporous silica-derived sol-gel glasses. The protein conformation can be then determined by circular dichroism spectroscopy (CD) to quantify properly folded proteins. We showed, for instance, that the hydrophobicity of the silica surface enhances the biological activity of the proteins. The utility of CD spectroscopy, based on left-handed and right-handed circularly polarized light absorption, has been reported in the literature to determine the protein secondary and tertiary structure in solution [36-38]. CD spectroscopy has proven to be a complementary tool to fluorescence and other spectroscopic techniques described in this chapter to relate the structure of the host matrix to the conformation of the protein [2,3,35,39].

Understanding the protein-folding process constitutes an endless challenge at this time, and the development of new approaches and techniques are necessary. The utilization of new technical approaches such as in-situ NMR spectroscopy [40] is considered to probe the important role of water on the protein folding in nanoporous silica-based sol-gel materials. The importance of functional groups such as fluorine to enhance the protein bioactivity has been demonstrated, but its role is not yet fully understood. The developments of new compatible materials associating different forces and functional groups (phosphate/fluorine) that enhance the protein folding are currently investigated by our group. A new patented technology recently developed and known as Fluoro-Raman spectroscopy, aka C-F Raman Spectroscopy [41], is introduced and represents a promising technique to be employed to assess the role of fluorine on the enhancement of the protein folding. The technique is currently employed for the determination different silica-protein interactions (hydrogen bonds, F-proteins interactions, for instance) with water as possible intermediate for new biocompatible sol-gel materials. Overall, this chapter will provide an overview on the preparation, applications, and characterizations of bioencapsulated materials that aim to enlighten the most important parameters affecting the protein folding in crowded environments in general and in particular, in nanoporous silica-based materials prepared from functionalized organosilane precursors.

Chapter 2

2. BIOENCAPSULATION VIA SOL-GEL PROCESS IN SILICA-BASED MATERIALS: METHOD, MATERIALS, BIOAPPLICATIONS, AND CHARACTERIZATION TECHNIQUES

2.1. Protein Bioencapsulation via Sol-Gel Process

The sol-gel process has been described in detail by Brinker and Scherrer [14]. Typically, alkoxide monomers (tetramethoxysilane, TMOS or tetraethyle thoxysilane, TEOS) undergo hydrolysis in the absence or presence of an acid or base catalyst to form silanols. The silanols link together through polycondensation to form a siloxane network characterizing the inorganic silica matrix. Residual alcohols (e.g., methanol in the case of TMOS) are then released upon polycondensation to form methanol or ethanol depending on the silica precursors used (TMOS or TEOS, respectively). The preparation of inorganic silica gels has been extended to novel hybrid nanomaterials in which organic and inorganic species are mixed at the molecular level [42]. The pioneering works on those materials were led by Schmidt [43] and Wilkes [44]. Organic-inorganic hybrids or ORMOSILs (ORganically MOdified SILicates) concern the modification of TMOS or TEOS by adding organically modified silanes $R_nSi(OCH_3)_{4-n}$, to undergo hydrolysis and polycondensation using the same protocol as described earlier for the unmodified TMOS glass and to obtain sol-gel matrices such as [(100-x) TMOS: (x) $R_nSi(OCH_3)_{4-n}$, n = 1, 2, 3, R = functional organic group, x = mol% composition] where -Si-O-Si- is the siloxane network in which some of the bridging oxygens of a SiO_4 tetrahedron are substituted by one, or several, organic groups.

These two classes of materials (unmodified and organically modified silica glasses) are good hosts for the incorporation using the sol-gel process of different additive such as pigments, organic dyes [45,46] metal particles and a variety of chemical compounds that can all be combined with the silica sol in solution before gelling and incorporated in silica-based organic-inorganic hybrid matrices for different applications purposes (e.g., photonics, optics and so on), depending notably on the properties and the functionality induced by the surface modification for instance [47-52]. The special physical and chemical properties of silica are quite attractive from the viewpoint of the compatibility with biomolecules and living cells. The thermodynamic stability of the Si–O bond is of 452 kJ/mol indicates a strong inertness that excludes interference with enzymes and functions typical of differentiated cells. The approach to sol-gel silica loaded with biological systems, particularly enzymes, has produced important and valuable implementations in biotechnology [53-62] and opened tremendous possibilities to understand the protein-folding process.

To achieve the entrapment of an active protein in sol-gel derived silica glass, it is necessary to maintain the active conformation of the protein within the matrix. Traditional routes to glass formation involve high temperature processes that are not compatible with the preservation of biomolecules or cells upon encapsulation, so these silica-based nanoporous glasses obtained via the sol-gel process are ideal for bio-immobilization (Figure 1). The mild conditions associated with the synthesis at ambient temperature make the sol-gel route offer new possibilities in biotechnology. During the sol gel process, after mixing the alkoxysilanes, a low concentration of acid (HCl ~ 10^{-2} N) is added in order to increase hydrolysis rates leading to the formation of fully hydrolyzed precursors. The hydrolysis can be performed by sonication, as it is well known that silanes do not solubilize easily in water. Upon sol formation and prior to gelation, a buffer solution is added to adjust the pH to values compatible with biomolecule viability (pH 7). After pH adjustment, a buffered enzyme solution is added [63]. The proteins are usually unstable outside a narrow pH range around pH 7. Therefore, they are kept in a buffered solution at neutral pH before addition. One drawback upon the hydrolysis of alkoxides is the formation of harmful alcohol byproducts. This problem can be partially overcome by evaporation or distillation of the alcohol from the hydrolyzed sol before addition of biomolecules [64]. The amount of water for hydrolysis of alkoxysilane is also another parameter to take into account the sol-gel process for bioencapsulation. It has a dramatic influence on gelation time [65]. For low-water content, an increase in the amount of hydrolysis water decreases the gelation time, though there is a dilution effect as well. It can be predicted that for higher water content, the gelation time increases with the increasing quantity of water. The molar ratio of silane to water was set to 2:3, respectively, for the experiments carried out in our laboratory to allow a slow polycondensation process. A wet-aging period of two weeks was necessary to obtain a transparent encapsulated sol-gel glass. Given these conditions, the biomolecule is entrapped within the growing oxide network [66]. To obtain a uniform thickness of glass samples, the hydrolyzed sol mixture containing the protein can be transferred to one-mm spacing PET cassette (Invitrogen, USA) and left to gelation at 4°C (Figure 1). This method is useful for the quantitative analysis of the protein structure via spectroscopic techniques such as CD spectroscopy, as we can obtain a constant thickness for all protein-encapsulated glass sheets, and we can then normalize easily the content of helical protein. It is worth noting that the measurements, for instance, are carried out on protein bio encapsulated wet-gels of which rectangular piece of glass of defined size fitting a two-mm tick CD quartz cell is immersed in a buffer solution (pH 7) for equilibration. Furthermore, this confinement method allows also a better diffusion of the biomolecules in the silica matrix independent of the thickness of the host material.

The transparency of bioencapsulated samples (Figure 1, inset) is very important parameter for the study of the protein conformation via spectroscopic techniques but also for the development of chemical and biochemical sensors that rely on changes in an absorbance or fluorescence signal. However, it is important to note that optically transparent hybrid nanoporous sol-gel glasses can be obtained only taking into account the molar composition of the silane precursors for some glass systems.

We showed recently that the samples with molar ratios of organically modified silane (organic groups being alkyl or fluorine) above 15 mol% to TMOS (85 mol%) lose their transparency and could not be studied by spectroscopic techniques [2].

The sol-gel matrices obtained before and after aging and drying processes can lead to several types of materials ("wet gels," xerogels or aerogels depending on the drying process, Figure 1).

Before drying, a liquid (buffer + protein) remains in the pores of the bioencapsulated nanoporous sol-gel glass that is immersed in an aqueous media (buffer at neutral pH), as seen earlier to probe the protein conformation using CD spectroscopy; the material is known as a "wet gel." If the liquid in the pore is removed by thermal evaporation (which is not ideal for protein bioencapsulation studies), pore shrinkage occurs (the pores collapse) and the monolith obtained is known as xerogel [67].

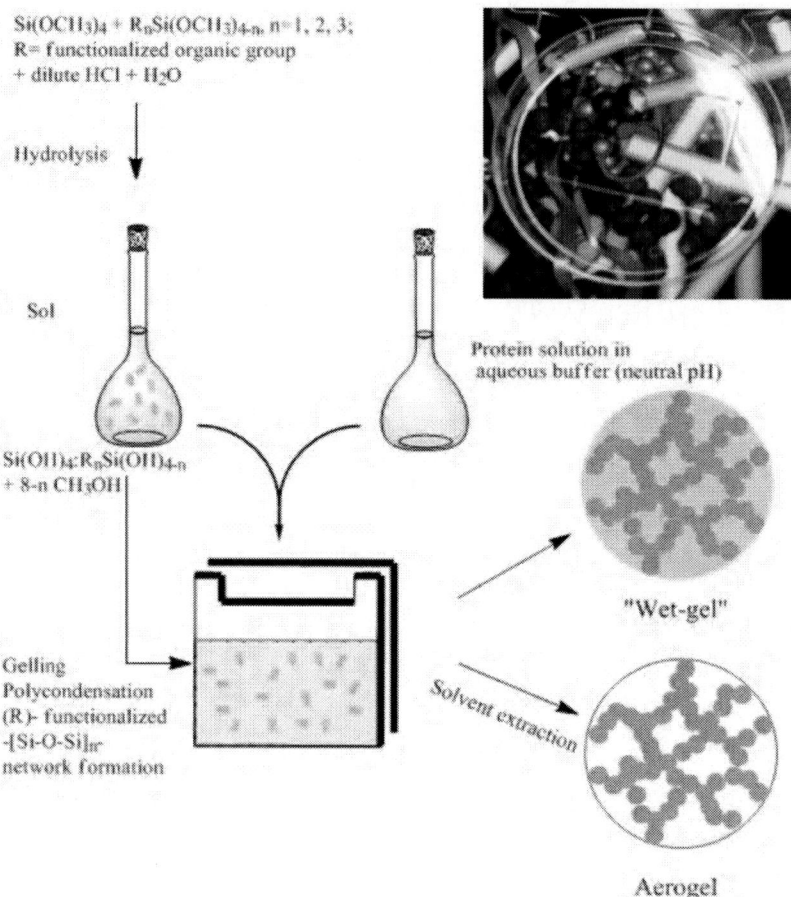

Figure 1. Schematic experimental process describing the protein encapsulation using the sol-gel process through hydrolysis/polycondensation of organosilane precursors leading to two possible materials after sol aging in a PET cassette at 4°C: (i) the "wet-gel" (the pores are filled with the protein in an aqueous buffer) and (ii) the solvent extraction from the "wet-gel" leads to aerogel materials characterizing the solid host matrix; the inset figure shows the transparent nanoporous protein bioencapsulated sol-gel glass (one-mm-thick sheet) in a petri-dish.

Aerogels of unmodified silica or hybrid organic-inorganic mesoporous silica-based materials can be obtained via solvents exchange or supercritical liquid CO_2 drying methods [68-70] at ambient temperature to remove the liquid from the pores and to preserve the porous structure and the functionality (given by the organic group) of the host matrix. This method is described later with the solvent exchange drying method (Part 2.3.1.). The cautious drying leading to aerogels materials has then the advantage of preserving the transparency, the original physical and structural properties of the host materials upon protein encapsulation without altering the structure and pores after drying but also to characterize the host matrix.

In general, the silica-based sol-gel matrix unmodified or modified with organic functional groups hosting the protein and prepared as described earlier, is composed of a rigid siloxane network typical of mesoporous silica-based materials. Their pores sizes are ranging from five to ten nm, and their surface areas are comprised between 600 and 1000 m^2/g. The size and chemical nature of the pores can be tailored to conform the biomolecule size and to avoid the protein or enzyme leaching. The modification of TMOS or TEOS surface with organoalkoxysilanes offers the possibilities of changing the properties of the materials (hydrophobicity, polarity and surface charges, pores size, steric effects). Changing the ratio of tetraalkoxysilane to organotrialkoxysilane can control the cation exchange capacity, polarity of porous surface, the hydrophobicity/hydrophilicity [71] and, consequently, the local environment of the protein. We can then associate the physical properties and structure of the host materials to the protein conformation in the "wet gels."

2.2. Materials and Bioapplications

Most of entrapped enzymes retain their catalytic activity and appear to be protected against denaturation [25]. A wide range of enzymes, antibodies and other biomolecules have been trapped within sol gel matrices and have been reported over the years [20-22, 54, 72-76]. In 1990, Avnir and co-workers [56] described the entrapment of proteins into alkoxysilane-derived silicate materials using the sol-gel method. This group demonstrated that a series of enzymes, including aspartase and alkaline phosphatase, could be entrapped into silica materials derived from tetraethyl orthosilicate (TEOS) with retention of enzymatic activity. The biomolecules such as atrazine chlorohydrolase [77], lipase [78], human serum albumin [79] entrapped in organically modified sol-gel materials show improved performances, including storage stability, excellent activity retention, etc. Dunn's and Zink's groups in 1992, demonstrated that other proteins, such as cytochrome *c* and myoglobin, could be entrapped into tetramethyl orthosilicate (TMOS) derived silicates with retention of O_2 binding ability [66]. More recently, the use of a silica precursor bearing a covalently attached sugar has been used for the formation of protein-doped sol-gel derived silica [80-81]. Silica materials containing covalently bound gluconamide moieties provide a biocompatible environment for entrapped enzymes, allowing long-term stability and reusability of the enzyme even after repeated washing steps. Polyols such as poly(glyceryl silicate) known as PGS have been prepared via the sol-gel method and represent a new class of materials that can allow the tuning of hydrophilicity and hydrophobicity [61]. The materials are highly biocompatible with water solubility and have the advantage of releasing a nonvolatile, bioprotective alcohol after hydrolysis, which would also function as a drying control additive (DCA).

The addition of synthetic or natural polymers such as poly(ethylene glycol) (PEG) to TEOS and organosilane-derived sol–gel material offer also enhanced material properties such as optical clarity and dehydration/rehydration stability, which result in significant improvement in the medium term stability of entrapped lipase as compared to entrapment in absence of polymer additives [82]. The use of polycationic polymers into ormosils materials could improve the performance of flavoproteins [83,84]. Ormosils and polymers are shown to be good to modulate enzyme activity. The ormosil methyltrimethoxysilane (MTMS) doped with chitosan shows good biocompatibility for immobilization of glucose oxidase (GOD) [85]. Horseradish peroxidase (HRP) and acetylcholinesterase can also be encapsulated in ormosils [86-88]. Taking into account the advantages of these enzymes and the advances in ormosil technology, they have been successfully employed in design of biosensors [89]. Indeed, these immobilized enzymes were evenly distributed in the sol–gel matrices and gave good response for biosensing. Glucose oxidase and hexokinase for instances, that are used as ex vivo biosensors, could find applications as in vivo biosensors, such as in the continuous monitoring of glucose in the management of diabetes [90-94].

The biomolecules entrapped in ormosil nanoparticles can be efficiently employed in drug delivery and other pharmaceutical and medical applications [11]. The organically modified and anticancer drug doped nanoparticles (diameter ca. 30 nm) obtained through aqueous dispersion have been used in photodynamic cancer treatment [12]. Water-insoluble photosensitizing anticancer drug, 2-devinyl-2-(1-hexyloxyethyl) pyropheophorbide, was entrapped in non-polar core of micelles by hydrolysis of triethoxyvinylsilane. The resulted nanoparticles were spherical, highly monodispersed, and stable in aqueous system.

An array of substances, including catalytic antibodies, DNA, RNA, bacteria, antigens, cells, and protozoa, have been encapsulated in silica-based sol-gel materials [95,96].

Porous siloxane-based organic inorganic materials have been employed, too, as heterogeneous catalysts for industrial applications [7, 19-22, 25]. In particular, the design of organically modified hydrophobic glasses for encapsulating lipases, a class of enzymes that mediate lipid reactions and function naturally at hydrophobic interfaces, has received much attention [23, 26-28, 31, 78, 97, 98] due to the potential application of these enzymes in the conversion of fats and oils to other products of value [99].

All these materials' applications are related to the protein function that is dependent on proper folding of the polypeptide chain into a specific, three-dimensional structure, and that protein structure, in turn, may be affected by its immediate surroundings. Thus, in the crowded and confined environment of a silica matrix, the physico-chemical properties of the glass are expected to influence the protein structure and activity. This hypothesis may be tested by the protein conformation into chemically modified silica networks and by monitoring slight alterations in protein conformation by spectroscopic techniques [36,39,99-101]. It is then important to characterize the conformation of encapsulated enzyme as function of the structure and properties of the host matrix. In such case, different techniques exist to be able to characterize the silica-based host materials and the conformation of the protein in the nanopores to understand the different parameters and the nature of the local microenvironments influencing the protein folding.

2.3. Probing the Silica-Protein Interactions and Protein-Folding

2.3.1. Characterization of the Silica Host Matrix

As explained earlier, the advantage of obtaining aerogels after encapsulation is to preserve the structure of the host matrix in order to be able to characterize them by physico-chemical techniques for which the characterization is only possible under an inert or dry atmosphere. It is then possible to determine the properties of the host materials after encapsulation of biomolecules or proteins, for instance, as we retrieve the original host structure of bioencapsulated materials. It has then possible to analyze the structure and surface properties of the host matrices via solid state NMR, Porosity and Brunauer Emmett Teller (BET) surface area measurements via N_2 adsorption, surface microscopy and other techniques of interest to determine the properties of the host matrix by performing a cautious drying process of the wet gels at ambient temperature. The physically adsorbed water is evacuated during a cautious drying process at ambient temperature to retain its original structure and to avoid capillary associated with drying shrinkage and consequently pore collapses. In order to strengthen the gel network, multi-step solvent exchanges were undertaken by washing of the gels with non polar solvents [68-70].

The gels were washed with acetone for three days (one time per day) and left aged in stand-closed glass vials in acetone during this time. Subsequently, the same process during the same time was applied with pentane (which has a low surface tension).

Once the solvent exchange was completed, the vials were opened and covered with aluminum foil and the drying fluid was evaporated at a rate limited by a one-mm diameter pinhole aperture to allow a slow evaporation of the pentane. The resulting solid materials were additionally dried for 24 hours at 37°C to form aerogel samples suitable for physical analysis. The drying process can also be carried out via supercritical liquid CO_2 [69].

Porosity and BET surface area measurements via N_2 adsorption give us information about the influence of the pore size and additional microstructure of the host matrix (Figure 2).

The different characterizations of the host matrices showed that a variety of chemical modifications based on the addition of organosilane leads to a controlled pore size and pore distribution, which allows small molecules and ions to diffuse into the matrix while large biomolecules remain trapped in the pores.

The matrix is then described as mesoporous gels (average pore diameter in the 5 – 10 nm range, Figure 2(b)) with surface area in the range of 500-1000 m^2/g. For instance, unmodified TMOS and organically modified glass shows typical physisorption isotherms (Figure 2(a)) of mesoporous silica materials modified with a hysteresis profile for the unmodified TMOS-derived glass, with or without protein encapsulation corresponding to type H3 (reference given by IUPAC classification) [103,104].

This classification is associated with adsorption into slit-shaped pores or into the space between parallel platelets. In contrast, organically modified silica samples (modified with alkylalkoxysilane for instance), the hysteresis of type H2 was found.

This profile is associated with adsorption into bottleneck-shaped pores, that is, pores for which the inner compartment is wider than the entrance. The influence of the shape on protein folding is discussed in Section 3.4.

Electron microscopy techniques such as Scanning Electron Microscopy (SEM) and Atomic Force Microscopy (AFM) relate the possible silica-protein interaction and the surface morphology induced by the protein encapsulation.

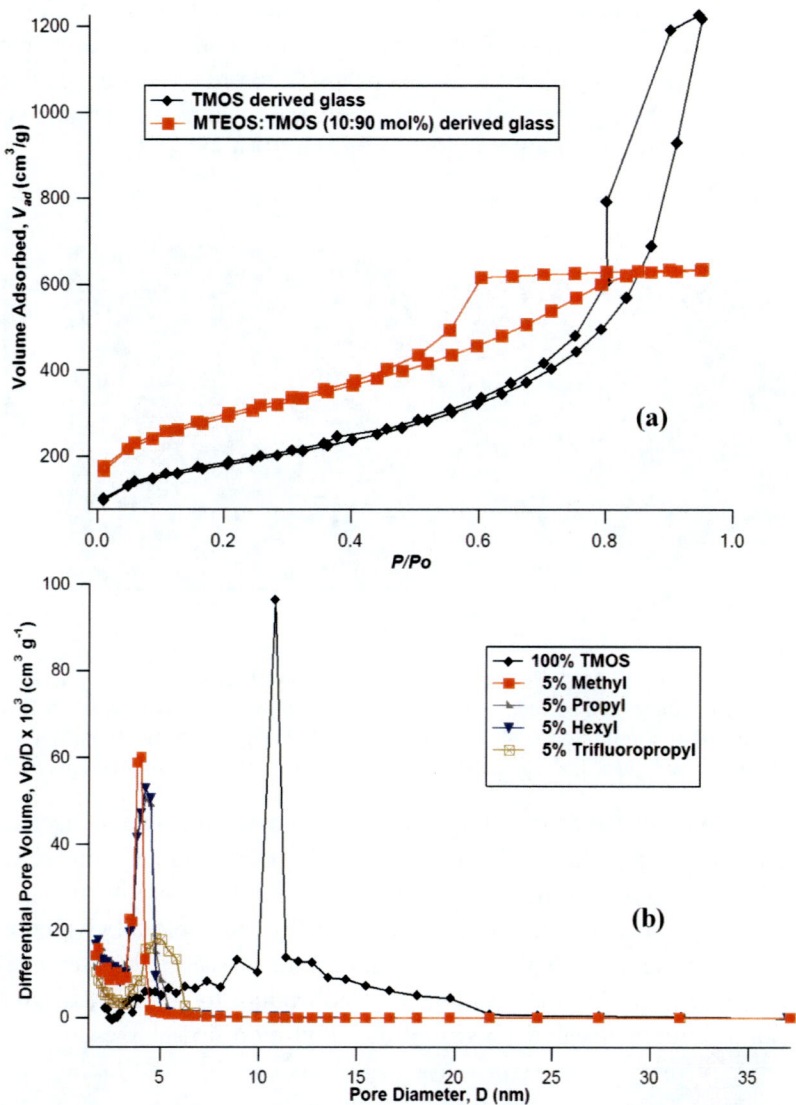

Figure 2. (a) N_2 adsorption-desorption isotherms for the unmodified TMOS and 10% methyl-modified glasses with apomyoglobin encapsulated, showing the two types of hysteresis profiles H2 and H3 characterizing the mesoporous materials in unmodified and organically modified; (b) pore-size distribution for different glass systems (unmodified TMOS and organically modified TMOS:RSi(OCH$_3$) [R= alkyl, fluoroalkyl] showing that the average pore diameter is about ten nm in the unmodified and four nm in the modified glasses, respectively [1-3].

SEM measurements of unmodified TMOS and alkyl-modified silica glass surfaces showed that spiral-shaped domains of interconnected silica particles forming TMOS were only observed upon encapsulation in the presence of apoMb. In the case of TMOS, the protein encapsulation resulted in largest domain structures and appeared to form a semi-array pattern (Figure 3(a)).

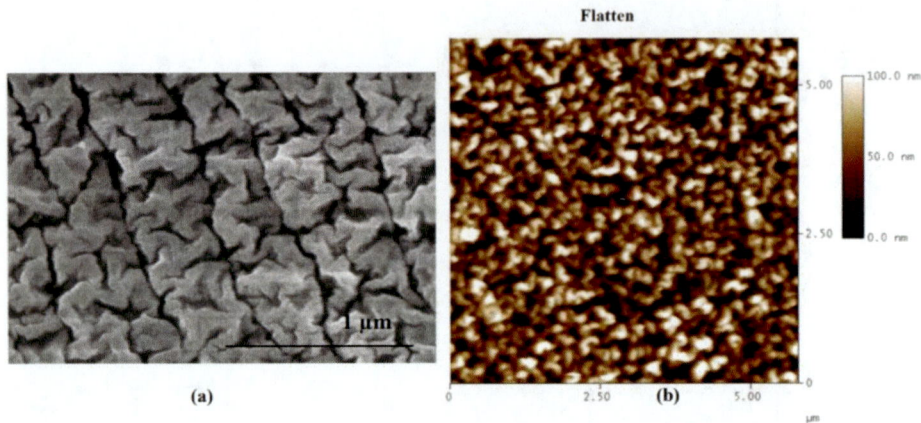

Figure 3. (a) Scanning Electron Microscopy image of apoMb encapsulated in unmodified silica glass with a patterned surface of semi-arrays consisting of serpentine interconnected silica particles; (b) Atomic Force Microscopy image showing 20 nm silica particles in organically modified glass TMOS:RSi(OCH$_3$) [R=Ethyl] with encapsulated apoMb [1-2].

We observed that the size of spiral-shaped domains of interconnected silica particles (Figure 3(b)) decreased at the surface of organic-inorganic hybrid sol-gel glasses due to the increasing hydrophobicity of the host matrix by modifying TMOS with organosilanes modifiers containing hydrophobic alkyl groups. The result suggested that apoMb may serve as a template or seed for agglomeration of the silica particles during the gelation process, leading to a particular patterning morphology. The smaller domain size of the alkyl-modified glasses may reflect a change in the strength of the silica-protein interactions that lead to the serpentine microstructure. SEM is then a useful technique to enlighten a change in the surface morphology of the host matrix induced by the protein-silica interactions, which depend also on the nature of the functional groups attached to the silica surface [2].

^{29}Si Magic Angle Spinning (MAS) NMR can be employed to determine the structure of the silica glass [31,79,98,104]. The method permits the identification of the silicon network units present in the different glass systems and the quantitative determination of siloxane (-Si-O-Si- linkages) and silanol (Si-OH) content in the glasses [35]. The spectra describe well the structure and silicon connections with distinctive peaks characterizing the units present in the different glass systems. The terminology of these condensations units Q^n, T^n, D^n, M^n are described according to Engelhardt et al. [105], in which Q denotes the four possible silicon-oxygen connections in TMOS, and T, D, M the three, two and one possible connections in organically modified TMOS glasses (n being the number of Si-O-Si linkages). The Q^n, T^n, D^n, M^n sets of peaks deconvoluted assuming Gaussian forms and surface area of each of the primary peaks lead to the determination of the overall cross-linkage (%), silanol –[Si-OH] content, and degree of condensation, which are useful pieces information that can be related to parameters influencing the protein conformation in the nanoporous silica glasses.

The thermogravimetry/mass and Fourier transform infrared (FTIR) spectra has been used to indicate that the entrapment of urease in a sol–gel silica film can form a silica network [106].

The characterization of sol–gel matrices include also small-angle X-ray scattering (SAXS), neutron scattering (SANS), light scattering (SALS), and fluorescence spectroscopy [107-109]. The application of SAXS to a number of gel systems has been reported by various authors [107-109] as the technique permitting the determination of a characteristic length of a particle and a fractal dimension, which give some information on the structure of the polymer (branched versus linear) and the growth mechanism. SANS can been applied to the study of silica sols [110-111] but can also be conducted on immobilized enzymes, as it yields structural information on complex biological systems in real time without damaging the structures involved [112].

2.3.2. Characterizing the Protein Folding in Nanoporous Sol-Gel Glasses

Spectroscopic techniques, in particular circular dichroism, fluorescence, and NMR, have provided crucial insights into the mechanism by which a polypeptide chain attains its native fold.

The encapsulation of myoglobin (Mb) in a functional conformation state has been demonstrated by Ellerby et al. and Lan et al. [66,10] who were among the first to enlighten it by changes in the absorption spectrum upon ligand binding. Saavedra et al. [113] encapsulated met-Mb and found that their encapsulation process resulted in perturbations of the Mb structure as evidenced in an altered absorption spectrum.

Spectroscopic characterization of the deoxy-HbA sol-gel-encapsulated samples via the visible resonance Raman and the UV resonance Raman spectra showed that TMOS sol-gel-encapsulated deoxy-HbA samples are very similar to that of the solution-phase deoxy-HbA, demonstrating that there is little perturbation of the tertiary and quaternary conformation of deoxy-HbA by the sol-gel environment. In addition, these spectroscopic studies in TMOS glasses indicate that encapsulated equilibrium populations of hemoglobins and myoglobins typically retain the same conformational distribution as in solution [114,115]. It has also been possible to trap specific quaternary states of the allosteric enzyme aspartate transcarbamoylase in TMOS silica matrix [116]. The confinement inhibits and slows quaternary conformational changes [117-119], while retaining catalytic activity by allowing the rapid transport of small molecules such as substrates and products [66].

It has been possible to measure the reaction kinetics of the isolated T and R states, as well as measuring the kinetics of the allosteric transition on a much slower time scale than in solution, using pyrene-labeled ATCase in hydrated sol gels. Previous studies on myoglobin and hemoglobin have clearly demonstrated that protein encapsulation influences the rates of conformational transitions [120], allowing trapping of otherwise unstable conformational states [121].

FRET (Forster Resonance Energy Transfer) is one also of these particularly useful techniques to study protein folding. The method, originally developed for trapping conformational intermediates of hemoglobin, has been applied to the study of the conformational dynamics of the native protein and the unfolding process of a mutant of the green fluorescent protein (GFP) both in buffer and in the presence of several concentrations of GdmCl, proving to be versatile and reliable [122,123]. Once included in the polymeric matrix (pore diameter *ca.* 10 nm), the protein molecules can be observed for a period of hours to days (and without the energetic features of the protein being affected by confinement) [101,124], allowing the monitoring of a complete (un)folding reaction. The ability of proteins to rotate in nanoporous sol-gel glasses can be measured directly using fluorescence anisotropy [79].

All spectroscopic [111-115], oxygen binding [114,117,125-127], and kinetic [117-120,128] studies indicate that the sol-gel encapsulation can be used to stabilize the conformation of initially encapsulated protein. However, the effect of encapsulation on local and global protein dynamics is strongly dependent on individual protein properties. Saavedra et al. [113] have characterized and studied by fluorescence spectroscopy the Bovine Serum Albumin (BSA) and myoglobin conformation upon sol gel bioencapsulation TMOS-based glass. They conclude that BSA, due to steric restriction, retains its native conformation, while apomyo globin (ApoMb) entrapment resulted in a significant loss of its native structure due to the fact that the host matrix may influence the thermodynamics of the protein and favor its instability. ApoMb is considered as soft protein and, therefore, is more susceptible to structural arrangement in the porous gel compared to BSA, which is considered as hard protein. This is another good reason to use apoMb in order to study the protein folding in nanoporous silica sol-gel glasses as function of its microlocal environment.

In the last decade, apoMb has attracted particular interest in both compu tational and experimental studies of protein folding because it provides important information about the intermediate state of protein folding. This model protein has been widely used for the study of protein-folding/unfolding process in solution [32-34]. The holoprotein (myoglobin) is very stable, but once its heme has been removed to form apoMb, the protein unfolds easily, and several partially folded states can be populated depending on its surrounding environment. From earlier experimental works in solution, it has been known that the apo-form of myoglobin shows three states under acid-induced unfolding with the formation of an intermediate at pH 4 [129-131]. Intensive NMR studies on apoMb have also revealed that the protein has a well-defined native form, and its structure is very similar to that of holomyoglobin [129,132,133]. In addition to these findings, it has been reported that the structure of the molten globule of apomyoglobin has a hydrophobic core associated with helices A, G, and H, whereas the B-helix is largely disordered [132]. An experiment for folding kinetics has shown that this molten globular state is on the folding pathway of apoMb [131]. Subsequently, many experimental works have been focused on early events of the folding to clarify what is the driving force for stabilizing the native and intermediate state of apoMb and for leading to a correct folding [131,134-137].

Recently, complementary studies on apoMb were performed to quantify the fraction of the properly folded protein (normalized per mass of protein) in nanoporous silica-based sol-gel glasses in order to associate the physical properties of the host matrix to the protein conformation. Many reports indicate that sol-gel glass encapsulation retains the activity and the native conformation of a wide variety of enzymes, but it is necessary to provide any insight on whether or not the glass composition also influences the enzyme structure.

We have reported previously [2], and this has been confirmed by Saavedra et al. [113], that apoMb, which is a protein of moderate stability in solution, becomes largely unfolded following encapsulation in unmodified TMOS silica glass using the sol-gel protocol described earlier. Consequently, surface effects on protein structure are a genuine concern for encapsulation studies. Organically modified TMOS glasses (organic groups being alkyl or fluoroalkyl groups) encapsulated with apoMb or RNase A were characterized by scanning electron microscopy, porosity and BET surface area measurements, ^{29}Si MAS NMR analysis, and circular dichroism spectroscopy (CD). These associated techniques have the advantages to monitor the protein conformation (secondary and tertiary structure) in account of the structure of surface properties of the host matrix. CD spectroscopy is based on left-handed and right-handed circularly polarized light absorption and has been reported in the literature for the determination of the protein secondary and tertiary structure in solution [36,37]. The measurement of the molar ellipticity of the protein as function of the wavelength in the far-UV region showed the presence of the signal at 222 nm, characterizing the secondary structure, whereas the tertiary structure of the protein was detected in the near-UV, which is the sign of its stability and biological activity. It is then easy to quantify the helicity of the protein using the following formula [38]: $C = \frac{\theta \times 100 \times MW}{[\theta]_{mol} \times d \times Na}$, where C is the concentration of the protein stock (mg/ml), $[\theta]_{mol}$ is the molar ellipticity (cm^2/dmol) for the protein at the considered wavelength, d is the path length of the cuvette (0.2 cm), MW is the protein molecular weight (g mol^{-1}), θ is the measured ellipticity (degrees), Na is the number of residues or the number of amino acids per protein ($Na = 153$ for ApoMb), the factor 100 originates from the conversion of the molar concentration to the dmol/cm^3 concentration unit, C is the protein concentration (mg/ml).

Chapter 3

3. PARAMETERS INFLUENCING THE PROTEIN CONFORMATION IN NANOPOROUS SILICA-BASED SOL-GEL GLASSES

3.1. INTRODUCTION TO THERMODYNAMICS— DRIVING FORCES AND INTERACTIONS INFLUENCING THE PROTEIN FOLDING IN SILICA-BASED NANOPOROUS MATERIALS

Recently, we focused on the influence of hydrophobic surfaces of nanoporous silica sol-gel glasses on protein folding that will be developed in the following part. We pointed out that proteins in the bioencapsulated silica gel are solvated by a layer of water. Since it has been shown that SiO_2 surface can induce order in water structure [138], we suggest that the water structure, which is a hydrogen-bonded network, is perturbed at a solid surface relative to its structure in the bulk solution [3,139]. We consider that the hydrophobic effect on folding equilibria is directly related to the formation of unfavorable water structure at the surface of the unfolded protein, and we suspect that in the studied case of organically unmodified glass (TMOS) that is formed of Si-O-Si linkages, the surface boundary induces a layer of water of high-free energy that is thermodynamically unfavorable. Therefore, the pores alter the average free energy of non-protein-associated (bulk) water and, consequently, the strength of the hydrophobic effect on protein folding [2,3]. Considering this effect, the modification of glass surface with hydrophobic organic groups can, therefore, influence the enhancement of the protein folding by altering the thermodynamically unfavorable water free energy.

We have to recognize that it is difficult to obtain thermodynamics data on the different interactions (such as surface hydration) induced by the microlocal environment change of the protein in nanoporous silica glasses compared to solution, the encapsulation of proteins (e.g., enzymes) in functionalized inorganic-organic nanoporous sol-gel glasses constitute an interesting way to manipulate the energy landscape described by Frauenfelder and co-workers [140-143], as it was shown that it can select specific conformations and/or increase the stability of functionally active states. The protein encapsulation can increase the enzyme stability by inhibiting via steric effects the large-scale conformational changes associated to expanded transition or denatured states [144-146]. The encapsulation can also induce the height of kinetic barriers' increase [114-116, 120, 125, 147-152] that will favor one conformation over others and by changing the relative depth of adjacent energy minima. In such case, this will alter the enzyme function [151, 152]. Most of thermodynamics studies have been carried out on protein in solution, as it is easier to employ the techniques such as DSC or calorimetry in the liquid phase. However, thermodynamics studies on the different parameters influencing the protein folding in the solid porous host matrix with the determination of water order and structure are currently undertaken by our group and represent an important challenge. Based on our recent studies, we will consider here the effects of surface hydration and hydrophobicity, solute effects with the presence of salts made from Hofmeister ions surrounding the protein in the bioencapsulated glasses. We will report the steric effects from silica glass and its surface modification with several organic groups attached to the silicon of organosilane precursors such as dimethyldimethoxysilane (DMDMS) or trimethylmethoxysilane (TMMS). We will compare the thermal stability of the protein in solution and confined in the porous host matrix. The influence of porosity and surface area of the glass have been studied to check if these parameters affect the conformation of the confined protein. We will suggest some possible interpretation based on thermodynamics considerations for some of these parameters and particularly for the surface hydration and hydrophobic interactions that seem to be the most important findings, explaining the enhancement of the protein folding in silica-based nanoporous sol-gel glasses.

3.2. The Surface Hydration and Hydrophobicity Influence the Protein Folding in Nanoporous Sol-Gel Glasses

The role of surface hydration and hydrophobic interactions are important in solution [159]. Water molecules influence specific interactions in cell biological systems, and yet it is extremely difficult to understand their effects in precise atomic models. Recent studies of water effects on biological macromolecules have made molecular biologists aware of the important role that this solvent plays in the structure and functions of proteins and other cell constituents. The water effects might also interfere thermodynamically in the protein-folding process in nanoporous silica materials. The proteins encapsulated in the silica gel, which can mimic the cell in a crowded environment, are solvated by a thin layer of water, and the removal of surface waters can destabilize the proteins by unfolding process. Water is the bridge to allow small solute molecules to be diffused from a bathing solution through the glass to the encapsulated protein. This diffusive connectivity between the bulk solvent and the gel interior is indeed consistent with a solvating layer. The solvating water resides in a limited space bounded on one side by the polar glass wall and on the other side by the surface of the protein. If we consider the surface hydration and the hydrophobic effect, then we have to describe them in terms of structural rearrangements of water molecules at the molecular interface. Thermodynamically speaking, hydrophobic interactions are very interesting [154,155], and the studies of small non-polar molecules showed that hydrophobic groups in water will tend to cluster together because of their mutual repulsion from water, not necessarily because they have any particular direct affinity for each other. The association of non-polar groups to form a "hydrophobic bond" is said to be "entropy driven," even though the effect is endothermic. In counterpart, the separation of two hydrophobic groups in water is an exothermic process. This exothermic effect is opposed by a significant and thermodynamically unfavorable reduction in entropy of the system, attributable to water structure rearrangements at the molecular interface. The origin of the hydrophobic effect can be understood by considering the entropy of hydration for non-polar molecules [156].

Recent simulations by Pande et al. [157,158] on peptides have shown that the water order has a strong influence on protein folding when they are trapped within carbon nanotubes. The decrease of the carbon nanotubes' diameters was found to thin and order the water layer that solvate the protein helix. Consequently, the loss of entropy in this water correlated exactly with the fraction of helicity lost in the peptide. The confinement in the spherical cavity favored helix formation when the solvent was allowed to equilibrate with the bulk, but disfavored helix formation when solvent was trapped within the cavity. Zhou [159-161] has recently proposed that water trapped within the cylindrical pore has a higher thermodynamic activity than bulk water and that hydrogen bonding between trapped water and the peptide backbone favors the coil/unfolded state. The results obtained recently on protein (apoMb) confinement in nanoporous silica glasses suggest also the intriguing possibility that the peptide is surrounded by an ordered water surface and responds by unfolding.

3.2.1. Hydrophobic Effects on Protein Conformation Induced by Silica Glass Surface Modification with Hydrophobic Organosilanes Precursors

In fact, we think that the ordering of water structure can modulate the strength of the hydrophobic effect experienced by the unfolded protein in the gel pore containing a polar hydrophilic surface, as this is the case for unmodified TMOS glass. Indeed, this hypothesis has been verified by encapsulating proteins in hydrophobic nanoporous silica glasses TMOS: $RSi(OMe)_3$ (R= hydrophobic organic group such as alkyl) [2]. The hydrophobic character of such glasses has been verified by several means, including contact angle measurements [162], solid-state ^{29}Si MAS NMR, and Fourier transform infrared spectroscopy [104]. Hydrophobic, organically modified silica glasses have been used previously to enhance the activity of glass-encapsulated lipases [23-29, 31]. Although most enzymes prefer an aqueous environment, it is rational to place lipases in a hydrophobic environment because these enzymes mediate reactions at lipid-water interfaces and because the substrates of these enzymes are highly non-polar molecules [23-29,31]. However, we can probe the change of protein conformation with apoMb, as it is easily unfolded and several partially folded states can be populated depending on its surrounding environment. So if a change in glass hydrophobicity alters both the fraction of properly folded enzyme and the partitioning of the enzyme substrate into the matrix, then we would discern easily one effect from the other based on CD analysis and the characterization of the host matrix properties depending on the silica surface functionalization. When apoMb is encapsulated in such hydrophobic organically modified glass systems with R= n-alkyl chain (n=1-6), we can compare with the protein conformation in the unmodified hydrophilic TMOS glass [2]. The hydrophobicity of the pore walls was tuned by the surface modification of TMOS-derived glass with organically modified silanes, and, surprisingly, we observed that relatively small amounts of alkyl-modifying precursors (5-15%) can have a drastic effect on the fraction of properly folded protein in a glass matrix. The folding effect becomes more favorable as the alkyl chain length increases from one to six carbons. The average structures of proteins encapsulated in the pores of wet-aged-based silica glasses were determined by CD spectroscopy. The CD signal at 222 nm characterizes the protein secondary structure, and, therefore, indicates its potential biological activity [36-38]. The signal originating from the protein in glass samples may be compared directly to the protein signal obtained in dilute solution because the data are normalized to units of molar ellipticity. For instance, Figure 5 illustrates this tendency. The CD spectra provided for the unmodified and modified glasses at a fixed value of 10% RSi(OMe)$_3$ show a significant increase in the secondary structure of apoMb observed in the following order: unmodified < methyl (MTMS) < n-propyl (PTMS) < n-butyl (BTMS) (Figure 4(a; b; d; e)). The molar ellipticity of apoMb in the far-UV approaches the spectrum (Figure 4(f)) of the native protein in dilute solution (in 10mM phosphate buffer, pH 7) as the alkyl chain length increases. In contrast, apoMb was found unfolded in the unmodified TMOS glass (Figure 4(a)). The 100% TMOS-derived glass shows a flat CD signal and yields the weakest molar ellipticity (4,300 deg cm^2 dmol^{-1}), indicating that apoMb is largely unfolded in the control TMOS matrix (Figure 2) [2,3,35].

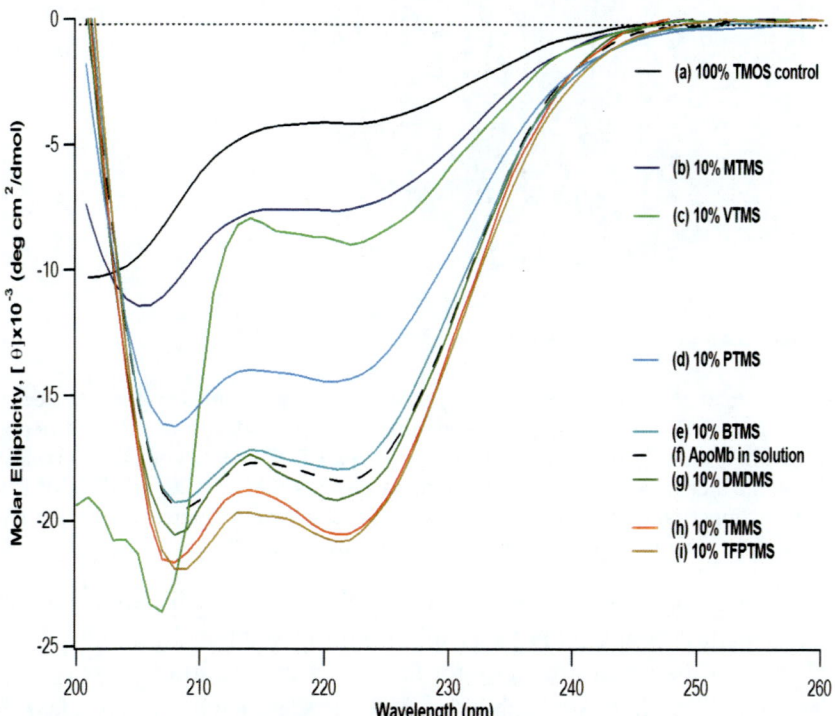

Figure 4. CD spectra in the far-UV region for apoMb in unmodified TMOS glass and modified glasses TMOS:R$_n$(OCH$_3$)$_{4-n}$ (solid curves) and in dilute solution (dashed curve) (*f*) All spectra were measured in the presence of potassium phosphate (10 mM, pH 7). The helicity of encapsulated apoMb increases with the hydrophobicity characterized by the increase of alkyl chain length (*b; d; e*) of the organic modifier R, its fluorine functionalization with trifluoropropyltrimethoxysilane (TFPTMS) (*i*) or the decrease of siloxane network n>1 with dimethyldimethoxysilane (DMDMS) and trimethylmethoxysilane (TMMS) (*b; g; h*); the spectra for apoMb encapsulated in the vinyl-functionalized silica glass (VTMS) (*c*) showed also that the protein folds [2,3,35].

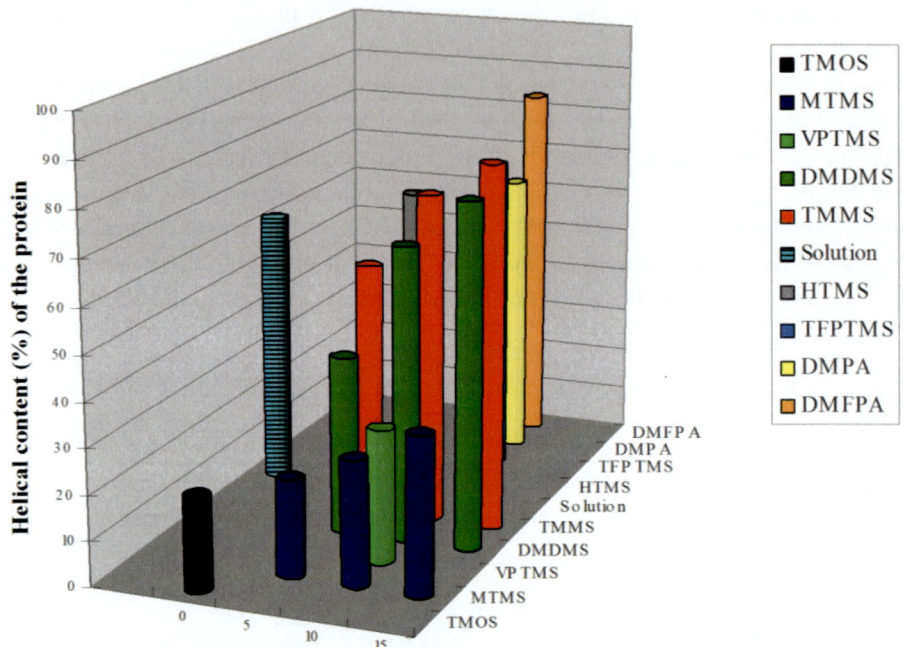

Figure 5. α-helical content of apomyoglobin in solution (potassium phosphate buffer (10 mM), pH=7), in different glass systems (unmodified TMOS and organically modified TMOS:$R_n(OCH_3)_{4-n}$) calculated from CD spectra using CDNN deconvolution software [162]. The helical content for some modified glass systems (MTMS, DMDMS, TMMS) is also shown as function of their molar composition to illustrate the increase of the helicity with the content of hydrophobic organic groups attached to the silicon.

At a wavelength of 222 nm, where molar ellipticity is an approximate measure of helical content, the value for apoMb in solution is -19,000 deg cm^2 dmol^{-1}, whereas the molar ellipticity for the 10% butyl-modified glass is -19,750 deg cm^2 dmol^{-1}, indicating slightly more helical structure in the glass than observed in solution.

By contrast, the relationship between protein structure and glass hydrophobicity is confirmed and strengthened by monitoring apoMb structure in a set of glasses of increasing molar content for each of the alkylsilane reagents employed in this study [2]. To have a better picture of the protein helical content in the different glass systems studied and their composition, the helical content of apoMb has been determined using the CDNN software algorithm from Böhm et al. (Figure 5) [163]. The secondary protein structure is composed of α-helices that have their importance in the biological activity and stability of the protein [163] but also to understand the protein-folding process. When the molar content of RSi(OMe)$_3$ is varied from 0-15 %, a corresponding increase in helical content of apoMb is observed for each set of glasses (e.g., methyl-modified glass (MTMS), Figure 5). This is surprising because apoMb is not expected to favor a hydrophobic environment; one would expect that a hydrophobic environment to favor the unfolded conformation of the protein because, once unfolded, the hydrophobic core of the folded state should prefer to interact with its hydrophobic surroundings as seen earlier. Consequently, these results contribute to strengthen the idea that the protein conformation might be directed by water structure rearrangement at the silica/protein interface.

3.2.2. Hydrophobic Effects Induced by the Decrease of siloxane – [O-Si-O]- Network Dimension by Glass Surface Modification with Multiple Hydrophobic Alkyl Groups Attached at the Silicon of Organosilane Precursors

Recent results [35] obtained with transparent glasses prepared from tetramethoxysilane (TMOS) and modified with a series of mono-, di- and tri-substituted alkoxysilanes, $R_nSi(OCH_3)_{4-n}$ (R= methyl-, n = 1; 2; 3) of different molar content (5; 10; 15%) to obtain the decrease of the siloxane linkage (-Si-O-Si-) showed the same trend but with more helical structure for the protein on account of the substitution with methyl groups (Figure 5). The molar ellipticity of the protein in the sol-gel glass has been determined by circular dichroism spectroscopy (CD) in the far-UV region as function of TMOS modification with methyltrimethoxysilane (MTMS), dimethyldimethoxysilane (DMDMS) and trimethylmethoxysilane (TMMS) and the molar composition (x) in the system [(100-x) TMOS: (x) $(CH_3)_nSi(OMe)_{4-n}$] system (n = 0; 1; 2; 3 and x = 0; 5; 10; 15 mol %) at a fixed value of 10 mol % of the modified silane $(CH_3)_nSi(OMe)_{4-n}$ (n = 1; 2; 3), a significant increase in the secondary structure of apoMb is observed upon TMOS glass modification in the order: methyl (n=1) < dimethyl (n=2) < trimethyl (n=3), Figure 4 (b; g; h). The protein transited from an unfolded state (Figure 4 (a)) in unmodified glass (TMOS) to a native-like helical state in the organically modified glasses. The molar ellipticity increases from 7,200 deg cm^2 dmol^{-1} (n = 1) to 21,520 deg cm^2 dmol^{-1} (n = 3). The organically modified glasses stabilize and tend to fold the protein in a rational manner with the methyl-Si substitution that features the decrease in the siloxane network. This controlled substitution with methyl groups confirms the relationship between the protein conformation and the silicon network structure of the glass. This trend tendency was also confirmed as function of the molar composition of the glass with the organically modified silane content (5% < 10% < 15%) for a given glass system (Figure 5). We observed that at fixed value of 5 mol % $(CH_3)_nSi(OMe)_{4-n}$ (n = 1; 2; 3) for each glass, the molar ellipticity of apoMb in the far-UV already approaches the spectrum of the native protein, as defined in dilute potassium phosphate (KPhos) solution at neutral pH (Figure 4(f)). In case of 10% molar content for dimethyl- (Figure 4(g)) and trimethyl-modified (Figure 4(e)) glass systems, apoMb reaches a higher molar ellipticity (-19,100 and -21,520 deg cm^2 dmol^{-1}, respectively), indicating that the protein is more helical in the glass compared to the native protein in the dilute solution. Moreover, the protein ellipticity of -24,000 deg cm^2 dmol^{-1} for 15% molar content of TMMS in the TMOS-modified glass was found even comparable to the stable folded heme-bound holoprotein in solution, which has a value near -24,000 deg. cm^2 dmol^{-1} at 222 nm. A hyperhelical structure has been reported for apoMb following encapsulation in trifluoropropylmodified glasses, but this effect was attributed to the presence of organic fluorine, and the helicity of the protein did not exceed that of the heme-bound holoprotein [3]. The α-helical conformation was found to represent about 89% in 15 mol % TMMS-modified glass while, the content of α-helices was about 32 % for same molar composition in the MTMS-modified glass (Figure 5) and while apoMb in solution (10 mM Kphos) has 62% of helical content. In fact, it is clear that the TMOS modification with an increasing content of terminal hydrophobic groups favors highly and even leads to the maximum helicity of the protein due to the presence of hydrophobic groups that change the water ordered structure at the silica/protein interface (Figure 6).

If we consider the structural characterization of the local environment for the protein in such glass systems, and to give an interpretation of this helical trend, we have recently analyzed the ^{29}Si MAS NMR spectra for the unmodified TMOS and 10 mol % MTMS, DMDMS and TMMS-modified glasses [35]. We observed that the cross-linkage decrements of about 3% at each methyl group attached at the silicon in the modified-glasses TMOS:$(CH_3)_nSi(OMe)_{4-n}$ (n = 1; 2; 3). Consequently, we have less -Si-O-Si- connections in the order 100 % TMOS > 10% MTMS > 10% DMDMS > 10% TMMS. On the other hand, in the same order, the silanol content -Si-OH, slightly increases of about 0.5% and consequently, the degree of condensation or condensation efficiency ([(Cross-linkage (%) / Cross-linkage + -(Si-OH) (%)] or ([number of siloxane chain] / [number of active oxygen = number of alkoxide]) decreases about 0.5%. The decrease of cross-linkage in the silicon network is then not due to the formation of Si-OH groups, but to the presence of -Si-(Me)$_n$ terminal groups.

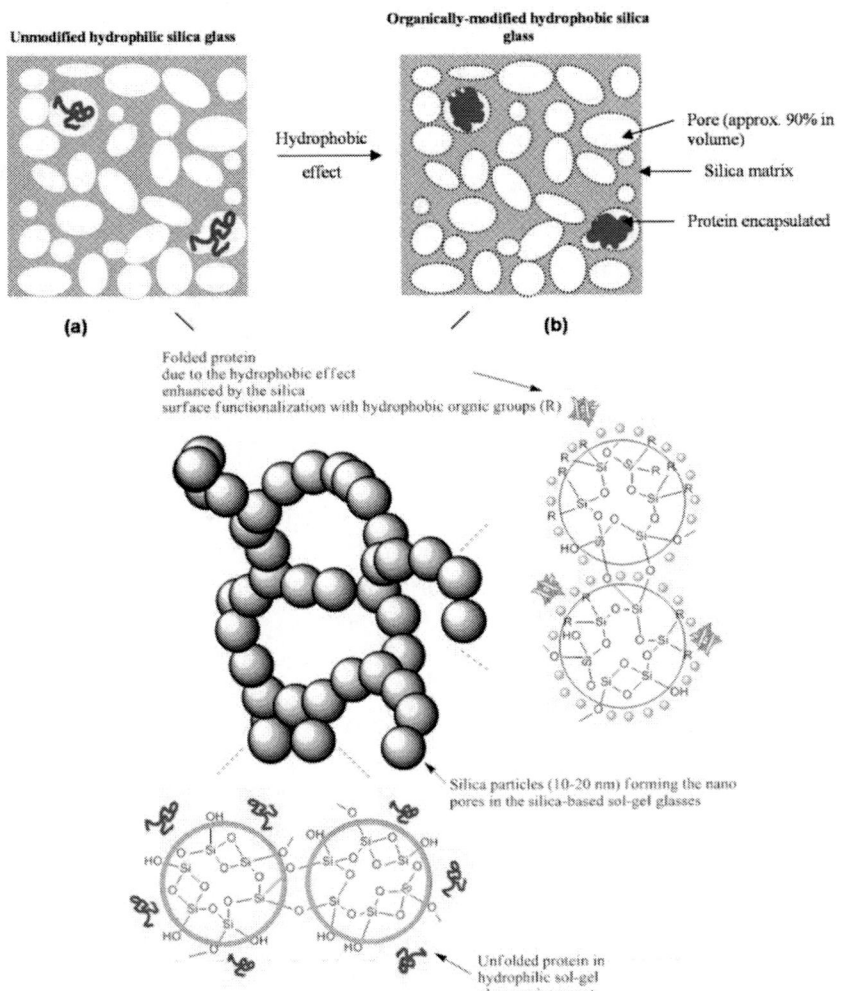

Figure 6. Protein encapsulated in silica-based nanoporous sol-gel matrix. The protein conformation depends on its surrounding environment and particularly on the interfacial water structure rearrangement that can be altered by the presence of hydrophobic organic groups attached at the silica surface. The square (a) represents the silica matrix containing hydrophilic pores such as in unmodified TMOS glass composed of Si-O-Si linkages; the ordered water molecules are represented around the pores in clear blue color. The protein is found to be unfolded. The square (b) represents the "hydrophobic glass," the hydrophobic effect is strengthened by the hydrophobic organic groups (R) of the silica matrix that alter the water structure arrangement, contributing to the folding of the protein. (c) Schematic representation of the silica particles forming the pores with hydrophilic or hydrophobic surfaces, representing the possible silica/protein interactions with water molecules as intermediate inducing a change in the protein conformation.

The change in the cross-linkage is then ascribed to the capping effect of the methyl groups. It is well known [165] that when we modify TMOS [Si(OMe)$_4$] with organically functionalized alkoxysilane RSi(OMe)$_3$ (R = methyl, for instance), the cross-linkage decreases due to the capping effect of the methyl. The small increase of Si-OH content and the consequent decrease of the condensation efficiency are due to the steric and capping effects of the methyl groups that hinder the formation of Si-O-Si connections in the order 100% TMOS > 10% MTMS > 10% DMDMS > 10% TMMS. In that context, we are tempted to conclude that the important parameter related to the protein helicity in these modified glasses is not the slight increase of the silanol but the larger decrease of the siloxane linkage –Si-O-Si- with the presence of hydrophobic Me-Si terminal groups characterizing the pores walls (Figure 6).

In fact, as described earlier, the surface hydration of the organically modified TMOS glass with hydrophobic groups at the silicon has a related effect favoring the protein helicity. The decrease of -Si-O-Si- connections is associated with less hydrogen bond interactions. In the unmodified glass TMOS, the silicon network of TMOS glass matrix contains a regularly spaced pattern of oxygen atoms that may act as hydrogen-bond acceptors [166]. The experiments have led to the hypothesis that apoMb is unfolded due to the altered properties of water in the glass. We have seen in the previous part that the silicon dioxide surface can induce order in water structure [138] and that it is highly probable that the water structure is perturbed at a solid surface relative to its structure in the bulk solution [139,167]. We suspect that in unmodified "polar" TMOS, the glass boundary induces a layer of water of high free energy (thermodynamically unfavorable). Water in these structures is more ordered than bulk water due to stronger hydrogen bonds and is, therefore, entropically unfavorable to protein folding. We can consider then that ordered water solvating a protein within a sol-gel reduces the hydrophobic effect in protein folding. In contrast, the functionalization of the glass surface with hydrophobic organic groups perturbs the water structure at the silica/protein interface and leads to the enhancement of the protein folding as illustrated in Figure 6.

Indeed, the hydrophobic effect on folding equilibrium is directly related to the formation of unfavorable water structure at the surface of the unfolded protein, and since this unfavorable water must partition to the protein surface from the bulk phase, it seems logical that the driving force for the hydrophobic effect is reduced by increasing the average free energy of the bulk water. In contrast, in the case of organically modified networks with hydrophobic groups, the pores are tailored by hydrophobic organic groups. Therefore, the pores alter the average free energy of non-protein-associated (bulk) water and, consequently, the strength of the hydrophobic effect on protein folding [3]. Experiments on protein adsorption to compare the protein conformation with bioencapsulated apoMb have led to interesting results about the role of water structure in protein folding for both approaches [3]. Adsorbed apoMb on TMOS modified glass with hexyltrimethoxysilane (5% mol) has shown that the protein has less than half the helical content found in solution, whereas the encapsulated protein exceeds the helical content found in solution. The protein was found more helical when encapsulated than when adsorbed with respect of the same glass system and composition. It should be noted, however, that hyperhelical conformations have also been reported for proteins adsorbed onto strongly hydrophobic surfaces, including Teflon [167-169] and trichloromethylsilane-coated quartz [170]. In fact, the unfolded conformation of the encapsulated protein in unmodified TMOS glass (which consists of Si-O-Si network) may be driven by desolvation at the silica surface that is entropically favorable, because silica behaves only as proton acceptor in forming bond with the solvent. The interfacial water molecules have then an ordered structure and are oriented in a similar manner with their protons pointing toward the oxygen atoms of the silica resulting in a low-entropy state. The disruption of this unfavorable water layer at the silica interface is generated by the introduction of hydrophobic alkyl chain by the decrease of the siloxane network dimension Si-Me for Si-O in organically modified glass that consequently enhances structure of neighboring protein molecules. If we consider the protein adsorption from the bulk solution, compared to the encapsulation in organically modified glasses during the sol-gel process, the protein docks with the silica surface at regions (locations having the highest free energy state of water) distant from the alkyl groups that disrupt the low-entropy hydration layer. Consequently, the adsorbed protein is prone to unfold on the glass surface to minimize the total number of thermodynamically unfavorable interactions between water and silica. This is in agreement with the general view that water, in general, controls the protein adsorption to surfaces [139].

3.2.3. Solute Effects and Hofmeister Ions Effects

In addition to this latest overview on the surface hydration and the hydrophobicity induced by the modification of the silica surface with organic groups and their effects on protein helicity in nanoporous sol-gel glasses, the utilization of solutes or salt solutions that are derived from Hofmeister ions are important and contribute to complement the study on the hydration effect at the silica/protein interface.

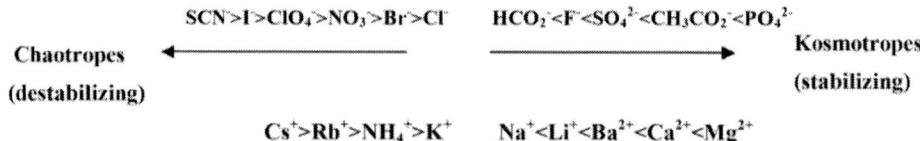

Figure 7. Hofmeister ions ranked as function of their degree of water structure forming and stabilizing. Chaotropes are known to disrupt water structure due to weak H_2O interactions and kosmotropes or stabilizers enhancing the water structure due to strong H_2O interactions.

It is known that ions classified as chaotropes have a destabilizing effect on enzymes [171] and aid in salting-in proteins in solution, while kosmotropes have a stabilizing effect on proteins and are effective salting-out agents (Figure 7) [172,173].

Chaotropes are often referred to as water structure-breakers, and kosmotropes are water structure-makers [174-177]. Furthermore, ion effects are generally additive, and anion effects tend to dominate solution behavior due to their asymmetric effect on polarizable water molecules [178]. Interesting information has been obtained when sol-gel silica based glasses were equilibrated with various 1.0 M salt solutions. A dramatic ion-specific increase in the helical content of encapsulated apoMb was observed. The secondary structure was found to be enhanced with the use of the following salt solutions in the order: $KH_2PO4 > N(CH_3)_4Cl > (NH_4)_2SO_4 > KCl > LiCl$. These same 1 M salt solutions had no significant effect on the native conformation of the soluble protein at pH 7.0 or on the molten globule-like conformation of the soluble protein at pH = 3.8. By contrast, in the presence of 1 M phosphate (pH 7), the helical content of encapsulated apoMb in unmodified TMOS glass approached that of the native protein in solution. In the case of KCl concentration, there was no effect on the secondary structure of encapsulated apomyoglobin. This effect may be explained by the fact that both K+ and Cl− ions fall in the middle of the Hofmeister series between kosmotropic ions and chaotropic ions [172,173]. Overall, these results suggest that water structure and protein hydration may be important determinants of apoMb conformation within the pores of the silica matrix.

Indeed, it was shown in a recent paper [2] that an increase in phosphate concentration from 10 mM to 1.0 M yields an increase in the molar ellipticity of apoMb in the hydrophylic unmodified TMOS glass from -4300 to -15,000 deg cm^2 $dmol^{-1}$ at 222 nm. Thus, the favorable effects of molar phosphate concentration for the hydrophilic glass such as TMOS on protein structure are demonstrated. When TMOS is organically modified with alkoxysilane $RSi(OMe)_3$ for (R = n-alkyl (methyl, ethyl, propyl, butyl, hexyl, for instance), and that alkyl-modified glasses were equilibrated in the 1 M potassium phosphate solution, the salt effect on apoMb helicity was more subtle and diminished with increasing chain length of the alkyl group. This suggests that the effects of altered glass hydrophobicity and molar salt addition to enhance the protein helicity are mediated by different mechanisms acting in the water structure rearrangement at the silica/protein interface. Example is given (Figure 8) for 10% methyl-modified and 10% butyl-modified glass for instance.

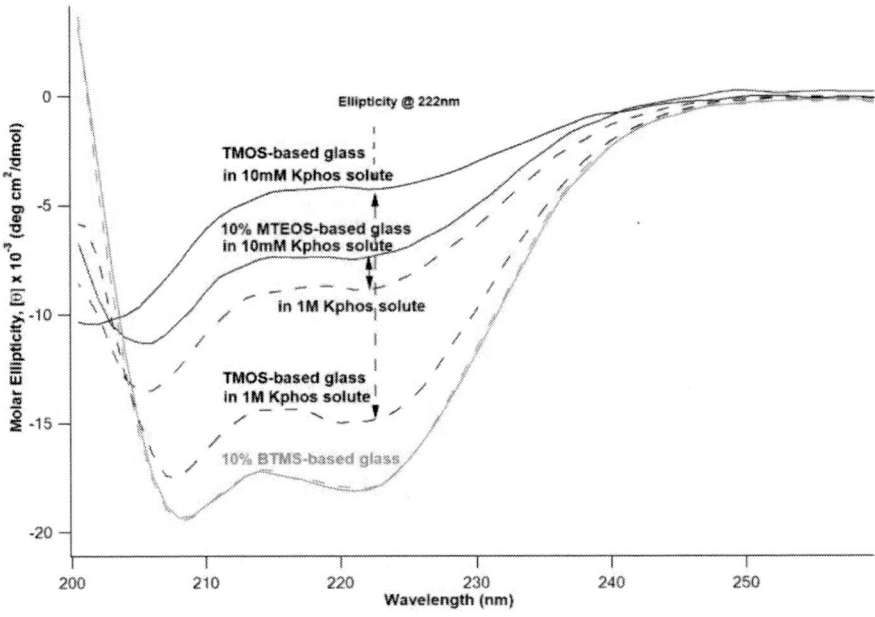

Figure 8. Influence of potassium phosphate concentration on apoMb structure in unmodified TMOS glass and modified glasses TMOS:$R_n(OCH_3)_{4-n}$. Each glass sample was equilibrated in 10 mM phosphate (solid curve) and in 1.0 M phosphate (dashed curve). The results show the effects of phosphate ions as water structure maker, which in high concentration alter the bulk water free energy, contributing to favor the hydrophobic effect and to enhancing the protein folding in hydrophilic organically unmodified TMOS glass. It is observed that the hydrophobic effect of the phosphate salt is less consequent in organically modified glass with the presence of one or several hydrophobic groups attached to the silica surface [2,35].

In fact, as described earlier, apoMb was unfolded in unmodified TMOS glass due to the "hydrophilicity destabilization." We have seen earlier that the water on the silica surface is ordered due to stronger hydrogen bonds, the hydrophobic effect is then diminished, and in terms of thermodynamics this effect is, therefore, entropically unfavorable. The interfacial water structure around the protein within the silica gel pores wall are thought to be oriented in a similar manner with their protons pointing toward the oxygen atoms of the silica. As a result, proteins unfolded within a gel pore can be stabilized and partially trapped in this conformation. The addition of high concentrations of Hofmeister salts, such as phosphate ion, presumably disrupts the gel-induced order in the water when the water reorganizes to solvate these ions, which, in turn, restores enough of the hydrophobic driving force to allow full folding of apoMb within the unmodified glasses. It seems then logical that in the case of organically modified glasses, the hydrophobicity the organic groups at the silica surface disturbed already the ordered water structure, and the effect of phosphate ions is less consequent on the protein folding.

3.3. STERIC EFFECTS INDUCED BY THE CHOICE OF CROWDED SILANE MODIFIERS IN TMOS- DERIVED SOL-GEL GLASSES THE HOST MATRIX

It is well known that cytoplasm of a living cell is a concentrated mixture of macromolecules (Figure 9) [4].

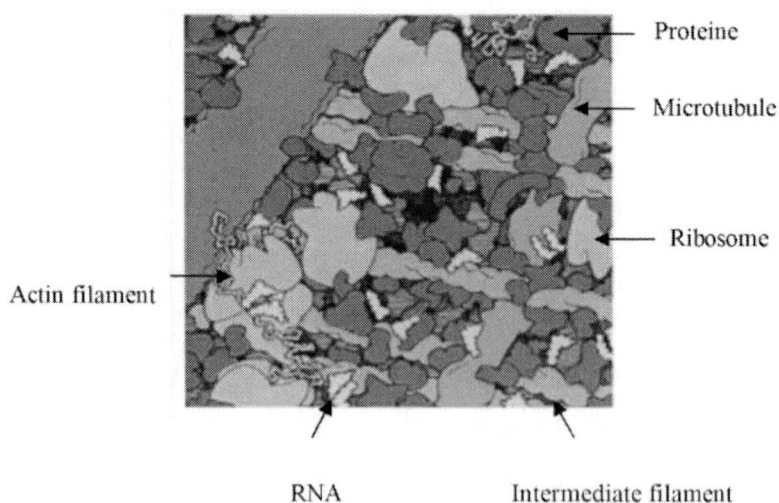

Figure 9. The crowded state of the cytoplasm in eukaryotic cells. Each edge of the square represents 100 nm in size. Small molecules have been omitted for clarity. Adapted from [5].

Membranes and elements of the cytoskeleton define local regions of confinement. The protein in nanoporous sol-gel glasses can be assimilated to that of a given macromolecule in a crowded and confined environment analogue to cytoplasmic membrane. In terms of thermodynamics activity in sol-gel glass or cytoplasm, it differs significantly from its activity in a dilute (ideal) solution. So the mimicking of the protein in such sol-gel medium is very useful. Thermodynamic studies with statistical models for crowding and confinement of macromolecules have been reported by Minton et al. [15] predicting that proteins will tend to favor compact or globular conformations induced by increasing crowding and confinement. The biological ramifications of macromolecular crowding have been considered in several reviews [5,6,15-17]. A number of reviews of both crowding and confinement effects have appeared during the past five years [179-185]. The presence of crowder influences equilibria between conformational states of a macromolecule by favoring conformations that exclude less volume to crowder. In the case of protein folding, unfolded conformations are more expanded, and, thus, crowding is expected to favor the native state.

The conformation changes can be obtained as function of the host matrix properties and as seen in our recent results for apomyoglobin encapsulated in systems where TMOS is modified with MTMS, DMDMS and TMMS and the molar composition (x) in the system $[(100-x)$ TMOS: (x) $(CH_3)_nSi(OMe)_{4-n}]$ system (n = 0; 1; 2; 3 and x = 0; 5; 10; 15 mol %) [35]. The helicity of the protein was found to be enhanced upon TMOS glass modification in the order: methyl (n=1) < dimethyl (n=2) < trimethyl (n=3). The sterical hindrance due to the capping effect of several methyl groups at the silicon has then its importance on protein folding and may have an additive effect on the stability of the protein secondary structure.

3.4. INFLUENCE OF THE PORE SIZE, PORE SHAPE AND SURFACE AREA OF THE SILICA-BASED HOST MATRIX ON PROTEIN FOLDING

It has been reported that the pore size may affect the protein dynamics [144]. The protein can also act as a structural template around which the gel network can develop and form a porous inorganic polymer cage. When drying occurs, this pore shrinks but conforms to the dimension of the dopant biomolecule, while the protein undergoes slight conformation changes as it adapts to the new micro environment furnished by the silica network.

We have recently addressed the possibility that the protein encapsulation (ApoMb) in organically modified glasses such as in TMOS:R(SiOCH$_3$)$_3$, R= n-alkyl (n =1 to 6) may be inducing a length-dependent change in the specific surface areas and porosities of both the modified and unmodified glasses, contributing to the observed increase in the fraction of properly folded protein [2]. In fact, we observed that the average pore size decreases from 10.8 nm (in unmodified TMOS glass) to about 4 nm (in alkyl-modified glasses) in the absence of apoMb, and the average pore size decreases from 8.9 nm to about 4 nm when apoMb is present, respectively (Figure 2(b)). All of the modified glasses, from methyl to hexyl, exhibit the same approximate pore size (~ 4 nm) with or without protein encapsulation. Considering the enhancement of the helicity with increasing alkyl chain length, the results showing that the pores size are similar in all alkyl-modified glasses regardless the alkyl chain length suggest that the pore size has no related effect on the protein structure. There is also no correlation between the surface area of the host matrix and the protein helicity. For the bioencapsulated TMOS glass, the surface area is of ca. 600 m^2/g, while it increases to ca. 900 m^2/g upon TMOS modification with R= n-alkyl groups (n=1-4), The surface area being similar for all the modified glasses, it does not seem to follow by itself a trend dependency to the enhancement of the protein helicity. These conclusions are also strengthened by the experiments carried out for the apoMb encapsulated in the system [(100-x) TMOS: (x) (R)$_n$Si(OMe)$_{4-n}$, R= CH$_3$] system (n = 0; 1; 2; 3 and x = 0; 5; 10; 15 mol %) [2,35].

It should be noted also that if one assumes that the average pore size of the bulk glass also reflects the average size of a protein-occupied pore, then one must conclude that glass-encapsulated apoMb inhabits an extremely confined environment; the calculated pore size is approximately 4 nm in all alkylmodified glasses, and the largest dimension of the protein is 4.2 nm from the crystal structure [186,187]. It should be noted that the actual pore size of a wet-aged glass before drying may be much larger than 4 nm. Thus, a molecule of apoMb should have ample room to rotate and alter conformation within the pores of a wet-aged glass.

In addition, physisorption isotherms (N$_2$ adsorption/desorption) on our bioencapsulated glasses (TMOS and alkyl-modified glasses) are typical of mesoporous silica materials [102]. It can be seen that the unmodified glass yields a different type of isotherm compared to the organically modified host matrix (Figure 2(a)). However, the isotherms obtained for the organically modified and unmodified silica glasses were found to be independent of protein addition. The hysteresis profile (Figure 2(a)) for the 100% TMOS-derived glass, with or without protein encapsulation, corresponds to type H3 by IUPAC [102,103]. This classification is associated with adsorption into slit-shaped pores or into the space between parallel platelets. In contrast, all of the modified silica samples exhibit type H2. This profile is associated with adsorption into bottleneck-shaped pores, that is, pores for which the inner compartment is wider than the entrance. The N_2 adsorption/desorption isotherm curves indicate a change in pore shape upon addition of an alkyl-modifying precursor, whether or not apoMb was present. Unfortunately, the difference in pore shape fails to explain the observed changes in protein structure because the same shape (hysteresis classification) was observed for all alkyl-modified glasses, independent of chain length.

The influence of pore size on the biological activity of enzymes in sol-gel matrices has also been a subject of discussion by other investigators. Reetz and co-workers noted that lipase activity in alkyl-modified silica host materials might be expected to increase with increasing pore volume due to better accessibility of the substrate, but the opposite was found experimentally [31]. The authors concluded that the catalytic activity of lipases is not related to the pore size but most likely a function of other factors related to the glass hydrophobicity, such as partitioning of the reactants or an interface-dependent conformational change in a specific loop that activates the enzyme. In a recent paper, Noureddini and Gao came to the same conclusion after finding no correlation between lipase activity and the specific surface areas of their alkyl-modified glasses [97]. In fact, even though nature did not design apoMb to function at a hydrophobic interface, apoMb structure was greatly enhanced in hydrophobic glasses providing direct evidence for the conformational change hypothesis that this is the main reason for the higher activity detected in modified glasses relative to unmodified glasses.

3.5. THERMAL STABILITY OF PROTEINS CONFINED IN THE POROUS HOST MATRIX

Several methods can be used for measuring protein thermal stability. Changes in tyrosine absorbance (at 280 nm), tryptophan fluorescence, and circular dichroism (CD) are all commonly used, as well as NMR, dynamic light scattering, and small-angle x-ray scattering are also used, though to a lesser degree, Recently, thermodynamic stability of proteins have also been measured microrheological techniques [188].

In our recent studies [2,35], we also employed CD spectroscopy to determine the thermal stability of apoMb in bioencapsulated glasses to monitor in-situ the conformational change in the glass as function of the temperature. When we substitute organically modified alkoxysilane RSi(OMe)$_3$ for TMOS in TMOS: RSi(OMe)$_3$ glass systems (R = n-alkyl), The results indicate that apoMb is extremely stable against thermal denaturation in both the modified and unmodified glasses when heated from 25 °C to 90 °C (Figure 10).

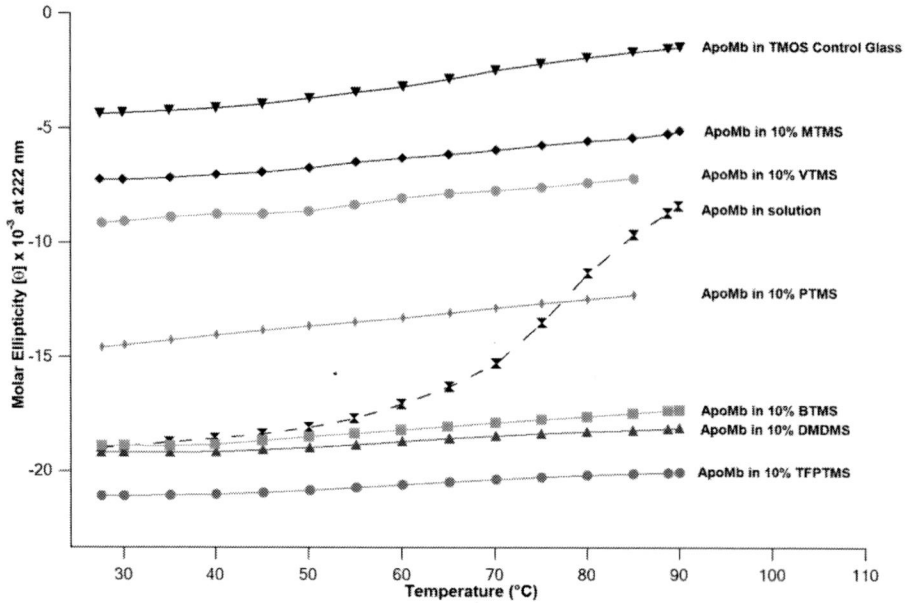

Figure 10. The thermal stability of the folded protein confined to different glass systems contrasts with the protein unfolding in dilute solution of potassium phosphate (10 mM, pH = 7) upon heating.

For example, the molar ellipticity of apoMb in the 10% butyl-modified glass only decreases from -18000 to -17300 deg cm^2 dmol^{-1} at a wavelength of 222 nm. At the other extreme, apoMb in solution proceeds to unfold through a broad transition from -19,000 to -8500 deg cm^2 dmol^{-1}. When a glass sample is held at 90°C for extended times (30 and 60 minutes), apoMb structure remains stable. The CD spectra of the 10% butyl-modified glass showed the features of a folded protein, even after a long period of heat treatment; the CD signal at 222 nm decreases only slightly from -18,000 to -17100 deg cm^2 dmol^{-1} after holding the sample at 90 °C for 60 minutes. The modest decrease in ellipticity may originate from a minor leaching problem at the elevated temperature. In contrast, the ellipticity of the protein in dilute solution decreases drastically after extended times at 90 °C, due to aggregation and precipitation of the destabilized protein. Upon heating from 25°C to 90°C, the thermal stability of apoMb entrapped in

10% DMDMS-modified glass [35], for instance, is also quite impressive, only a slight decrease in ellipticity was observed at a wavelength of 222 nm (Figure 10). This suggests that independent of the silicon network, the protein retains its helicity and stability against thermal denaturation due to the confinement of the protein in nanoporous glass. The outstanding thermal stability of proteins in organically modified glasses, as demonstrated for apoMb, bodes well for the use of such biomaterials in industry, including the development of new nanobiodevices [189]. At lower temperature consideration, it is also biologically important for organisms subjected to heat, dehydration or other environmental stress.

Chapter 4

4. ENHANCING THE PROTEIN FOLDING BY INTRODUCING AND ASSOCIATING HYDROPHOBIC AND STERIC EFFECTS IN MODIFIED SILICA-BASED POROUS GLASSES

4.1. INCORPORATING FLUORO-BASED ORGANOSILANES IN TO FORM SUPERHYDROPHOBIC CROWDED ORGANICALLY MODIFIED SILICA BASED HOST MATRICES

It has been shown that incorporating fluorine in the design of artificial amino acids has become prominent strategy to enhance the structural stability of peptides and proteins [190,191]. The decrease in polarizability of the carbon framework and the increase of the surface area upon fluorination increases the hydrophobicity of a molecule [192]. Consequently, the incorporation of fluoroalkyl groups can significantly increase diffusion across membranes such as the blood–brain barrier [193] and improve the pharmacokinetical properties of peptide-based drugs. C-F bond is highly dipolar, which enables its participation in polar interactions [194]. The strong inductive effect of the fluorine's atom combined with its high electronegativity affects the acidity and basicity of nearby functional groups [195]. Consequently, fluorination of alkyl groups can induce polarization of C-H bonds in proximity to the site of fluorine substitution. Besides the question of fluorine capability to participate in hydrogen bonds [196], some other important questions on fluorinated alkyl groups remain to be answered for the application of fluorine in protein-folding processes, and biotechnology in general [197]. One is the influence of the steric hindrance induced by the fluorine; the trifluoromethyl group range from twice the bulk of methyl [198] to close to the size of phenyl and tert-butyl [199]. Recent reports suggest that CF_3 better mimics bulkier alkyl groups rather than methyl in protein-binding pockets [200]. Another question may be concerned, indeed, with the influence of fluorine's inductive effect on the polarity of the C-F bond as well as nearby C-H bonds in a hydrophobic protein environment.

In that context, the protein encapsulated in fluorinated silica matrix such as TMOS:F_3C-R-Si(OCH$_3$)$_3$ (R= methylene or other functional groups) glass systems that can be used as soft host material to mimic the protein environment in the human body is of great importance. Recently, the increase of ellipticity and folding of apoMb encapsulated or adsorbed in fluorinated sol-gel glasses compared to the non-fluorinated ones have been enlightened using CD Spectroscopy with signals observable at 222 nm, characterizing the secondary structure at the far UV [3]. The results showed that a fluorinated silica matrix based on trifluoropropyltrimethoxysilane acting as host for the proteins has an effect on the protein folding. This was compared with the encapsulation of apoMb in a non-fluorinated silica matrix based on propyltrimethoxysilane, which difference is only the absence of the fluorine atoms. The experimental conditions and molar ratio of the precursors were similar. The results demonstrated the fluorine effect on the protein folding to near is native state.

Following our statements on the role of fluorine, its incorporation to new sol-gel glass systems might bring the dual steric and hydrophobic effect to enhance the protein folding. Understanding the role of the fluorine effect on the protein enhancement is investigated by our group at Fluorotronics, Inc. (San Diego, CA, USA).

4.2. INCORPORATING PHOSPHONATE GROUPS IN HYDROPHOBIC SILICA NETWORK

As seen previously, the kosmotropic effect of phosphate ions stabilize the water structure leading to local hydrophobic effect on the pores surface that affect the protein folding. In addition, fluoro-based nanoporous hybrid host materials have an influence on the ellipticity of the protein. Thus, by effects association, we recently enlightened the high helical content of ApoMb in silica-based fluorophosphonate materials [201]. The materials can be then suitable for the development of biosensors, bioreactors and biocatalysts. The idea to use difluoro methanediphosphonic acid (DFMDPA) as a precursor is very interesting if grafted to organically modified silica, but also if doped directly. Indeed, the local hydrophobicity can easily induced by the dual properties induced by the presence of the fluoro groups and the presence of the phosphate groups.

The CD spectrum (Figure 11) for 10% modified TMOS with DFMDPA goes in that sense.

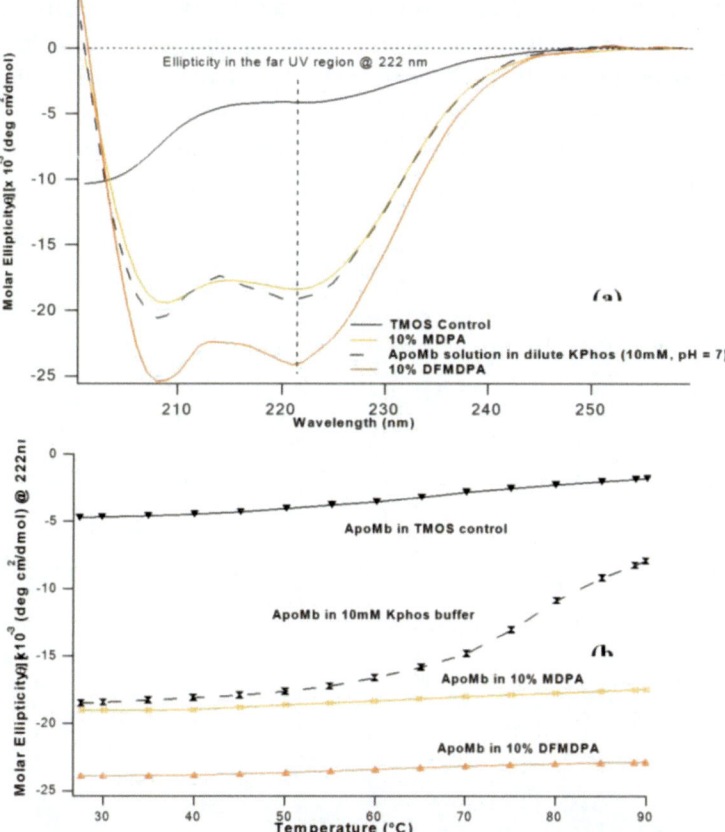

Figure 11. (a) CD spectra in the far-UV region for apoMb encapsulated in unmodified TMOS glass, modified TMOS glasses (solid curves) with 10 mol % of methylenediphosphonic acid (MDPA) and difluoromethylenediphosphonic acid (DFMDPA) and in dilute solution (dashed curve). All spectra were measured in the presence of potassium phosphate (10 mM, pH 7). The helicity of encapsulated apoMb increases with the hydrophobicity characterized by the dual properties of phosphate groups and fluorine; (b) Thermal stability of the folded protein confined to the different glass systems is also confirmed [201].

The presence of fluorine and phosphonate groups together enhances the protein helicity. These results are evident if we compare the protein-helical content in the unmodified glass TMOS but also in the modified glass with 10% methylenediphosphonic acid (MDPA) characterizing the absence of fluorine. The spectrum for MDPA-modified glass illustrates the important helical content (65%) of the protein that is slightly larger than in solution in 10mM phosphate buffer (62%) and much more than in the 10% methyl-modified glass (28%), suggesting that the presence of phosphonate groups enhances also the helicity of the protein (the different helical content are shown Figure 5). The addition of fluorine with DFMDPA increases the helical content of apomyoglobin (83%) to favor a hyperhelical structure. The helical content is even much larger than in 10% trifluoropropyl-modified silica glass (72% helical content). The materials show also a good thermal stability.

Chapter 5

5. EMERGING TECHNIQUES FOR A BETTER UNDERSTANDING OF PROTEIN INTERACTIONS AND CONFORMATIONS IN NANOPOROUS SOL-GEL GLASSES

5.1. IN-SITU MAS NMR

If we consider that water plays a major role as an intermediate between silica functionalized surface and protein folding, an important fundamental question in nanoscience is to know whether molecules in confined nanoporous spaces exhibit properties that differ significantly from those in the bulk [202-204], and it is essential that appropriate techniques are available to answer this question. We suggested earlier that interfacial water structure (at the silica surface) is more ordered than in the bulk water due to high hydrogen bonds induced by the silica surface and interfacial water. The result is that the ordered structure is thought to be entropically unfavorable to protein folding. However, this idea needs to be confirmed experimentally.

A key topic in this regard is the study of water molecules in nanoporous spaces that mimic crowded environment of the proteins [202]. Microcalorimetry can be used as a technical approach in solution, but it is rather difficult in a crowded and confined environment. In situ, solid-state NMR studies of adsorption processes can be used to map the structural evolution of water adsorbed in nanoporous voids within a widely used siliceous host material known as MCM-41 [205,206], which contains hydrophilic nanopores of typical diameter ca. 3 nm and has been used in a range of applications, including catalysis or drug delivery system. Although solid-state NMR is used widely in chemistry and materials science, its adaptation for in situ studies of adsorption processes is associated with specific technical challenges, both because the sample is located in a confined and relatively inaccessible space inside the NMR magnet and because high-resolution solid-state NMR usually involves rapid sample rotation. Although progress has been made [207,208] in the development of solid-state NMR techniques for in-situ studies of adsorption, a new experimental strategy exists with the specific feature of enabling the very earliest stages of the adsorption process to be probed directly. In this strategy, the species to be adsorbed is sealed in a glass capillary and inserted into the solid-state NMR rotor together with the host material on which the adsorption is to take place. The adsorption process is triggered by magic angle spinning (MAS). The NMR spectrum may be recorded immediately at the start of the adsorption process, and the time resolution then depends solely on the time required to record an individual NMR spectrum.

5.2. FLUORO-RAMAN SPECTROSCOPY

The introduction of functional groups favoring the crowding and hydrophobicity of the host matrix to enhance the protein folding need to be further investigated in encapsulated nanoporous sol-gel glasses. Recently, we developed a new patented fluoro-Raman spectroscopy (FRS) method via its device PLIRFA™ (www.fluorotronics.com) [41]. The technique is based on the high specificity and sensitivity, safety, fast data acquisition, and non-destructive detection of the C-F (carbon-fluorine) bond, which is rarely found in nature, in the spectral area of 550 cm^{-1} and 950 cm^{-1}. The C-F bond is now used as a Raman molecular label and presents many advantages. Indeed, the C-F bond is directly proportional to the concentration of the analyte. The features of the Fluoro-Raman spectroscopy allow us to use it for several purposes, including biological samples analysis and solid state materials. It can mainly detect, characterize and quantify the organo- or non-fluorinated compounds present in different types of materials such as solids (glass, quartz, polymers etc.), solutions (blood etc.) or even micro-organisms. Thus, our new technology can routinely be used to detect quickly and specifically fluorinated molecules (e.g. amino-acids, proteins and nucleic acids) in biological samples in order to understand the molecular mechanisms, propose a diagnostic, trace the molecules (e.g. imaging) or even screen patients treated with chemotherapy (e.g. 5-FU) [209].

Fluoro-Raman spectroscopy is currently used by our group as an effective tool to determine the role of fluorine on the protein folding. The utilization of this innovative technology has an important and fundamental impact in the deter mination of the different interactions (hydrogen bonds, etc.) via the different Raman signals induced between a protein and a solid sol-gel organic-inorganic silica-based host matrix prepared from fluoro-based organosilane precursors for intance. We expect the detection of possible differences in Raman signals for C-F bonds in the silica matrix itself compared to the fluorinated matrix with the protein in it. The results will be thus extremely useful for us to explain the role of the fluorine in the increase of the protein ellipticity.

CONCLUSION

This chapter enlightens the importance of functionalized organosilane precursors for the preparation of silica-based sol-gel glasses employed for bioencapsulation, which illustrate deeply their properties, performance and applications. The nature and choice of organosilanes can modulate the properties and the functionality of the biomaterials (hydrophobicity, crowding effects, and so on). The biocompatibility of the silica-based materials makes them suitable for the study of the protein-folding process or for the development of new bionanodevices. The materials obtained can be ascribed as "wet gel" when filled with water molecules that play a major role in the protein folding. They are highly porous (mesopores) materials and had the ability to adsorb easily solutes [2010,211], such as phosphate ions that can also play a role in enhancing the folding of the proteins. The materials are obtained at room temperature, which is convenient to host the proteins; they are synthesized using the sol gel route via hydrolysis and polycondensation from alkoxysilanes precursors with addition of the aqueous sol composed of the acid catalyst HCl and distilled water. The addition of alcohols as organic solvents is not necessary for the hydrolysis and polycondensation of the silica network because of the sensitivity and possible denaturation of the protein. Our current research focuses on the design of organic-inorganic hybrid sol-gel glasses for immobilization of proteins via surface engineering in order to mimic the ideal conditions to understand the proteins folding process for bioapplications. Indeed, we can tune the properties of the hybrid matrix leading to a change in the conformation of the protein simply by the choice and the nature of the Si-substituted organic group. In that context, the encapsulation of proteins in sol-gel glass or hybrid silica host matrix offers tremendous possibilities due to easy modifications of conditions (molecular crowding, hydrophobic effects, porosity etc.) that can be operated via the sol-gel method. It is important to note that the protein unfolding is responsible for many diseases (Alzheimer's, Diabetes's and so on). So the comprehension of the folding mechanism of proteins is fundamental from a scientific aspect. Finding the ideal experimental conditions and understanding the protein folding will also have a big impact on industrial applications for the development of new biosensors, bioreactors, and biocatalysts in which enzymes and proteins have to show biological, catalytic activities and, consequently, the ideal folding configuration. The implication of the sol-gel study is a verification of the assumption that the sol-gel encapsulation does not perturb the native structure of the protein in any significant way, thereby making the results from the sol-gel study applicable to the interpretation of solution-based studies. Indeed, it is known that sol-gel encapsulated proteins are fully solvated since the gel allows for the free and rapid diffusion of many water-soluble molecules and for the retention of water-dependent protein activity.

Encapsulating proteins such as apoMb, which is a model protein characterized by the removal of the heme group of myoglobin that makes it less stable and easily unfolded, has allowed us to quantify the properly folded protein in nanoporous silica glass systems hydrophilic unmodified (TMOS) or hydrophobic organically modified TMOS glasses with fluorine, n-alkyl groups, or fluorophosphonates. The structural analysis and surface properties of the host matrix, together with spectroscopic analysis (thanks to the optical quality and transparency of the glass) using in particular CD analysis to characterize the conformation of the protein within the host matrice, enlightened the important role and influence of the hydrophobicity induced by the surface modification of the host matrix. Our recent studies demonstrated that the decrease of the Si-O-Si network linkage associated with the capping effect of the hydrophobic groups, incorporated by modification of the silica surface in the porous organically modified silica-glass, enhance the protein helicity and its thermal stability. This modification induces a change in the surface hydration and consequently affects the silica/protein interactions. These results suggest also that the functionalized pores surface with the hydrophobic groups alter the average free energy of non-protein-associated (bulk) water and, consequently, the strength of the hydrophobic effect on protein folding. The choice of the silane precursor is very important; silanes bearing unsaturated functional groups such as vinyl are also very important for knowing if the proteins are still active; Kato et al. [212] reported the preparation and catalytic performance of lipases in vinyltriethoxysilane modified TMOS sol-gel glass. They reported that that lipases have higher hydrolysis activity and catalytic activity than other silane for 2-octyl acetate, TMOS, MTEOS and PTMS; our results on apoMb to monitor its conformation by CD confirms that the protein remains stable and has a helical structure between

MTEOS and PTMS. One important thing concerning VPTMS that our group is currently investigating is the use of the functional group to induce postfunctionalization while the protein is encapsulated and the monitoring in-situ of the protein folding as function of the environmental change.

Associating all the parameters that contribute to increase of the hydrophobicity (incorporation of "kosmotropic"-type groups that disturb the free energy and structure of water, fluorine group that is hydrophobic and which constitute a steric and crowding group) can lead to efficient bioactive silica-based materials where properly folded proteins are entrapped. In parallel, the development of new technique methods such as NMR to monitor the structural change of water structure in protein-encapsulated nanoporous silica-based materials and the new patented Fluoro-Raman technology to deeply study the influence of organic groups such as fluorine or the role of hydrogen bonds for specific host materials are currently employed by our group. Based on the parameters influencing the protein structure as function of its microlocal environment, the results will give us more information about how to control the microlocal environment to enhance the protein helicity. This could also lead to the design of highly efficient silica-based bioactive glasses using, for instance, superhydrophobic fluorine-based organosilane precursors.

REFERENCES

[1] Menaa, B.; Menaa, F.; Aiolfi-Guimaraes, C.; Sharts, O. Silica-based nanoporous sol-gel glasses: from bioencapsulation to protein-folding studies. *Int. J. Nanotech.* 2010, 7, 1-45.
[2] Menaa, B.; Herrero, M.; Rives, V.; Lavrenko, M.; Eggers. D. K. Favourable influence of hydrophobic surfaces on protein structure in porous organically modified silica glasses. *Biomaterials.* 2008, 29, 2710– 2718.
[3] Menaa, B.; Torres, C.; Herrero, M.; Rives, V.; Gilbert, A. R. W.; Eggers, D. K. Protein adsorption onto organically modified silica glass leads to a different structure than sol-gel encapsulation. *Biophys. J.* 2008, 95, L51-L53.
[4] Goodsell, D. S. The Machinery of Life. Springer-Verlag: New York, USA, 1998; 160pp.
[5] Ellis, R. J. Macromolecular crowding: obvious but underappreciated. Trends Biochem. *Sciences.* 2001, 26, 597-604.
[6] Ellis, R. J. Macromolecular crowding : an important but neglected aspect of the intracellular environment. *Curr. Opin. Struct. Biol.* 2001, 11, 114-119.
[7] Gupta, R.; Chaudhury, N. K. Entrapment of biomolecules in sol–gel matrix for applications in biosensors: problems and future prospects. *Biosens. Bioelectron.* 2007, 22, 2387–2399.
[8] Hungerford, G.; Rei, A.; Ferreira, M. I. C.; Suhling, K.; Tregidgo, C. Diffusion in a sol–gel derived medium with a view toward biosensor applications. *J. Phys. Chem.* B 2007, 111, 3558–3562.
[9] Tsai, H.; Doong, R. Preparation and characterization of urease-encapsulated biosensors in poly(vinyl alcohol)-modified silica sol–gel materials. *Biosens. Bioelectron.* 2007, 23, 66–73.
[10] Lan, E. H.; Dave, B. C.; Fukuto, J. M.; Dunn, B.; Zink, J. I.; Valentine, J. S. Synthesis of sol–gel encapsulated heme proteins with chemical sensing properties. *J. Mater. Chem.* 1999, 9, 45–53.

[11] Jain, T. K.; Roy, I.; De, T. K.; Maitra, A. Nanometer silica particles encapsulating active compounds: a novel ceramic drug carrier. *J. Am. Chem. Soc.* 1998, 120, 11092–11095.

[12] Roy, I.; Ohulchanskyy, T.; Pudavar, H. E.; Bergey, E. J., Oseroff, A. R.; Morgan, J.; Dougherty, T. J.; Prasad, P. N. Ceramic-based nanoparticles entrapping water-insoluble photosensitizing anticancer drugs: a novel drug-carrier system for photodynamic therapy. *J. Am. Chem. Soc.* 2003, 125, 7860–7865.

[13] Murphy, R. M.; Kendrick, B. S.; Chiti, F.; Dobson, C. M. Protein misfolding, functional amyloid, and human disease. *Annu. Rev. Biochem.* 2006, 75, 333-366.

[14] Brinker, C. J.; Scherer G. W. Sol–gel science: the physics and chemistry of sol–gel processing; Academic Press: San Diego, CA, USA, 1990.

[15] Minton, A. P. The influence of macromolecular crowding and macromolecular confinement on biochemical reactions in physiological media. *J. Biol. Chem.* 2001, 276, 10577–10580.

[16] van der Berg, B.; Ellis, R. J.; Dobson, C. M. Effects of macromolecular crowding on protein folding and aggregation. *EMBO J.* 1999, 6927-6933.

[17] van der Berg, B.; Wain, R.; Dobson, C. M.; Ellis, R.J. Macromolecular crowdings perturbs protein refolding kinetics : implications for folding inside the cell. *EMBO J.* 2000, 19, 3870-3875.

[18] Perham, M.; Stagg, L.; Wittung-Stafshede, P. Macromolecular crowding increases structural content of folded proteins. *FEBS Lett.* 2007, 581, 5065–5069.

[19] Bettati, S.; Pioselli B.; Campanini, B.; Viappiani, C.; Mozzarelli, A. Protein-doped nanoporous silica gel. In *Encyclopedia of Nanoscience and Nanotechnology;* Nalwa, H. S., Eds.; American Scientific: Stevenson Ranch, CA, USA; 2004, Vol. 9; pp 81-103.

[20] Avnir, D.; Coradin, T.; Lev, O.; Livage, J. Recent bio-applications of sol–gel materials. *J. Mater. Chem.* 2006, 16, 1013–1030.

[21] Gill I.; Ballesteros, A. Bioencapsulation within synthetic polymers (Part 1): sol–gel encapsulated biologicals. *Trends Biotechnol.* 2000, 18, 282–96.

[22] Jin, W.; Brennan J. D. Properties and applications of proteins encapsulated within sol–gel derived materials. *Anal. Chim. Acta.* 2002, 461, 1–36.

[23] Reetz M. T.; Tielmann, P.; Wiesenhofer, W.; Konen, W.; Zonta, A. Second generation sol–gel encapsulated lipases: robust heterogeneous biocatalysts. *Adv. Synth. Catal.* 2003, 345, 717–728.

[24] El Rassy, H.; Maury, S.; Buisson, P.; Pierre, A. C. Hydrophobic silica aerogel–lipase biocatalysts. Possible interactions between the enzyme and the gel. *J. Non-Cryst Solids.* 2004, 350, 23–30.
[25] Pierre, A. C. The sol–gel encapsulation of enzymes. *Biocatal. Biotransform.* 2004, 22, 145–170.
[26] Badjic´, J. D.; Kadnikova, E. N.; Kostic, N. M. Enantioselective aminolysis of an a-chloroester catalyzed by candida cylindracea lipase encapsulated in sol–gel silica glass. *Org. Lett.* 2001, 3, 2025–2028.
[27] Noureddini, H.; Gao, X.; Joshi, S.; Wagner, P. R. Immobilization of pseudomonas cepacia lipase by sol–gel entrapment and its application in the hydrolysis of soybean oil. *J. Am. Oil Chem. Soc.* 2002, 79, 33–40.
[28] Reetz, M. T.; Wenkel, R.; Avnir, D. Entrapment of lipases in hydrophobic sol–gel materials: efficient heterogeneous biocatalysts in aqueous medium. *Synthesis.* 2000, 6, 781–783.
[29] Kunkova, G.; Szilva, J.; Hetflejs, S.; Sabata, S. Catalysis in organic solvents with lipase immobilized by sol–gel technique. *J. Sol–Gel Sci. Technol.* 2003, 26, 1183–1187.
[30] Kim, K.; Park, J. K.; Kim, H. K. Preparation of nano-porous silica aerogel and its application to a bio-conversion process. *Resour. Process.* 2006, 53, 3–5.
[31] Reetz, M. T.; Zonta, A; Simpelkamp, J.; Rufinska, A.; Tesche B. Characterization of hydrophobic sol–gel materials containing entrapped lipases. *J. Sol–Gel Sci. Technol.* 1996, 7, 35–45.
[32] Choi, H. S.; Huh, J.; Jo, W. H. A novel water-soluble and self-doped conducting polyaniline graft polymer. *Biophys. J.* 2003, 85, 1492-1502.
[33] Barrick, D.; Baldwin, R.L. The molten globule intermediate of apomyoglobin and the process of protein folding. *Protein Sci.* 1993, 2, 869-876.
[34] Gulotta, M.; Rogatsky, E.; Callender, R. H.; Dyer, R.B. Primary folding dynamics of sperm whale apomyoglobin: core formation. *Biophys. J.* 2003, 84, 1909-1918.
[35] Menaa, B; Miyagawa, Y.; Takahashi, M.; Herrero, M.; Rives, V. Bioencapsulation of apomyoglobin in nanoporous organosilica sol-gel glasses: influence of the siloxane network on the conformation and stability of a model protein. *Biopolymers.* 2009, 91, 895-906.
[36] Greenfield, N. J. Using circular dichroism spectra to estimate protein secondary structure. *Nat. Protocol.* 2006, 1, 2876-2890.

[37] Greenfield, N. J. Determination of the folding of proteins as a function of denaturates, osmolytes, or ligands using circular dichroism. *Nat. Protocol.* 2006, 1, 2733-2741.
[38] Schmid, F. X. Protein Structure: A Practical Approach; Creighton, T.E. ed.; IRL Press: Oxford, UK, 1997; pp. 251-260.
[39] Shibayama, N. FEBS lett. Circular dichroism study on the early folding events of Beta-lactoglobulin entrapped in wet silica gels. 2008, 2668-2672.
[40] Matsuda, K.; Hibi, T.; Kadowaki, H.; Kataura, H.; Maniwa, Y. Water thermodynamics inside single-wall carbon nanotubes: NMR observations. *Phys. Rev. B* 2006, 74, 073415/1-073415/4.
[41] Sharts, C. M.; Sharts, O.; Gorelik, V. S. Method and apparatus for determination of carbon-halogen compounds and applications thereof. 2002, *US* 6, 445 449 B1.
[42] Sanchez, C.; Ribot, F. Design of hybrid organic-inorganic materials synthesized via sol-gel chemistry. *New J. Chem.* 1994, 18, 1007-1047.
[43] Schmidt, H. New type of non-crystalline solids between inorganic and organic materials. *J. Non-Cryst. Solids.* 1985, 73, 681– 691.
[44] Huang, H.; Orler, B.; Wilkes, G. L. Ceramers: hybrid materials incorporating polymeric/oligomeric species with inorganic glasses by a sol gel process. 2. Effect of acid content on the final properties. *Polym. Bull.* 1985, 14, 557–564.
[45] Menaa, B.; Takahashi, M; Tokuda, Y.; Yoko, T. Dispersion and photoluminescence of free-metal phtalocyanine doped in sol-gel polyphenylsiloxane glass films. *J. Photochem. Photobiol. A: Chemistry.* 2008, 194, 362-366.
[46] Menaa, B.; Takahashi, M; Tokuda, Y.; Yoko, T. High dispersion and fluorescence of anthracene doped in polyphenylsiloxane films. *J. Sol-Gel Sci. Technol.* 2006, 39, 185-194.
[47] Dunn, B.; Zink, J. I. Optical properties of sol-gel glasses doped with organic molecules. *J. Mater. Chem.*1991, 1, 903-913.
[48] Avnir, D. Organic chemistry within ceramic matrixes: doped sol-gel materials. *Acc. Chem. Res.* 1995, 28, 328-334.
[49] Avnir, D.; Klein, L. C.; Levy, D.; Schubert, U.; Wojcik, A. B. Organo-silica sol-gel materials. In the Chemistry of Organic Silicon Compounds; Rappoport, Z.; Apeloig, Y., J., Eds. Wiley and Sons: New York, USA, 1998; Vol. 2, pp. 2317-2362.
[50] Lev, O.; Wu, Z.; Bharathi, S.; Glezer, V.; Modestov, A.; Gun, J.; Rabinovich, L.; Sampath, A. Sol-gel materials in electrochemistry. *Chem. Mater.* 1997, 9, 2354-2375.

[51] Collinson, M. M. Recent trends in analytical applications of organically modified silicate materials. *Trends Anal. Chem.* 2002, 1, 30-38.
[52] Reisfeld, R. Prospects of sol-gel technology towards luminescent materials. *Opt. Mater.* 2001, 16, 1-7.
[53] Avnir, D.; Ottolenghi, M.; Braun, S.; Zusman, R. Doped sol–gel glasses for obtaining chemical interactions. *US Pat.* 1994, 5, 292,801.
[54] Avnir, D.; Braun, S.; Lev, O.; Ottololenghi, M. Enzymes and other proteins entrapped in sol-gel materials. *Chem. Mater.* 1994, 6, 1605-1514.
[55] Livage, J. Bioactivity in sol-gel glasses. *C. R. Acad. Sci.* Paris, 1996, 322, 417-427.
[56] Braun, S.; Rappoport, S.; Zusman, R.; Avnir, D.; Ottololenghi, M. Biochemically active sol-gel glasses: the trapping of enzymes. *Mater. Lett.* 1990, 10, 1-5.
[57] Rao, M. S.; Dave, B. C. Thermally regulated molecular selectivity of organosilica sol-gels. *J. Am. Chem. Soc.* 2003, 125, 11826-11827.
[58] Livage, J.; Roux, C.; Da Costa, J.M.; Desportes, I.; Quinson J.F. Immunoassays in sol-gel matrixes. *J. Sol–Gel Sci. Technol.* 1996, 7, 45-51.
[59] Dunn, B.; Zink, J. I. Probes of pore environment and molecular matrix interactions in sol-gel materials. *Chem. Mater.* 1997, 9, 2280-2291.
[60] Dave, B. C.; Dunn, B.; Valentine, J. S.; Zink. Sol-gel encapsulation methods for biosensors. *J. Anal. Chem.* 1994, 66, 1120A-1127A.
[61] Gill, I.; Ballesteros, A. Encapsulation of biologicals within silicate, siloxane, and hybrid sol-gel polymers: an efficient and generic approach. *J. Am. Chem. Soc.* 1998, 120, 8587-8598.
[62] Avnir, D.; Braun, S. Biochemical Aspects of Sol–Gel Science and Technology; Kluwer Academic Publishers: Amsterdam, Netherlands,1996; 143pp.
[63] Avnir, D.; Kaufman, V. R. Alcohol is an unnecessary additive in the silicon alkoxide sol-gel process. *J. Non Cryst. Solids* 1987, 192,180-182
[64] Ferrer, M. L; Del Monte, F.; Levy, D. A novel and simple alcohol-free sol-gel route for encapsulation of labile proteins. *Chem. Mater.* 2002, 14, 3619-3621.
[65] Sakka, S. Sol–gel synthesis of glasses: present and future. *Bull. Am. Ceram. Soc.* 1985, 64,1463–1473.
[66] Ellerby, L. M.; Nishida, C. R.; Nishida, F.; Yamanaka, S. A.; Dunn, B.; Valentine, J. S.; Zink, J. I. Encapsulation of proteins in transparent porous silicate glasses prepared by the sol-gel method. *Science* 1992, 255, 1113-1115.

[67] Pandey, P. C. A review on ormosil-based biomaterials and their application in sensor design. *J. Indian Inst. Sci.* 1999, 79, 415–430.

[68] Deshpande, R.; Hua, D. W.; Smith, D. M.; Brinker, J. C. Pore structure evolution in silica gel during aging/drying. III. Effects of surface tension. *J. Non-Cryst. Solids* 1992, 144, 32–44.

[69] Harreld, J. H.; Dong, W.; Dunn, B. Ambient pressure synthesis of aerogel-like vanadium oxide and molybdenum oxide. *Mater. Res. Bull.* 1998, 33, 561–367.

[70] Hwang, S. W.; Jung, H. .; Hyun, S. H.; Ahn, Y. S. Effective preparation of crack-free silica aerogels via ambient drying. *J. Sol–Gel Sci. Technol.* 2007, 41, 139–146.

[71] Schmidt, H. New type of non-crystalline solids between inorganic and organic materials. *J. Non-Cryst. Solids* 1985, 73, 681– 691.

[72] Lin, J.; Brown, C. W. Sol-gel glass as a matrix for chemical and biochemical sensing. *Trends Anal. Chem.* 1997, 16, 200-211.

[73] Livage, J.; Coradin, T.; Roux, C. Encapsulation of biomolecules in silica gels. 2001, *J Phys. Condens. Matter.* 13, R673-R691.

[74] Lei, C.; Shin, Y.; Liu, J.; Ackerman, E. J. Entrapping enzyme in a functionalized nanoporous support. *J. Am. Chem. Soc.* 2002, 124, 11242–11243.

[75] Lei, C.; Shin, Y.; Magnuson, J. K.; Fryxell, G.; Lasure, L. L.; Elliott, D. C.; et al. Characterization of functionalized nanoporous supports for protein confinement. *Nanotechnology* 2006, 17, 5531–5538.

[76] Besanger, T. R.; Brennan, J. D. Entrapment of membrane proteins in sol-gel derived silica. *J. Sol-Gel Sci. Techn.* 2006, 40, 209-225.

[77] Kauffmann, C.; Mandelbaum, R.T. Entrapment of atrazine chlorohydrolase in sol–gel glass matrix. *J. Biotechnol.* 1998, 62169–62176.

[78] Reetz, M. T.; Zonta, A.; Simpelkamp, J. Efficient heterogeneous biocatalysts by entrapment of lipases in hydrophobic sol–gel materials. *Angew. Chem. Int. Ed. Engl.* 1995, 34, 301–303.

[79] Brennan, J. D.; Hartman, J. S.; Ilnicki, E. I.; Rakic, M. Fluorescence and NMR characterization and biomolecule entrapment studies of sol–gel-derived organic–inorganic composite materials formed by sonication of precursors. *Chem. Mater.* (1999) 11, 1853–1864.

[80] Cruz-Aguado, J. A.; Chen, Y.; Zhang, Z.; Elowe, N.H.; Brook, M.A.; Brennan, J.D. Ultrasensitive ATP detection using firely luciferase entrapped in sugar-modified sol-gel derived silica. *J. Am. Chem. Soc.* 2004, 126, 6878-6879.

[81] Cruz-Aguado, J. A.; Chen, Y.; Zhang, Z.; Brook, M. A.; Brennan, J. D. Entrapment of Src protein tyrosine kinase in sugar-modified silica. *Anal. Chem.* 2004, 76, 4182-4188.

[82] Keeling-Tucker, T.; Rakic, M.; Spong, C.; Brennan, J. D. Controlling the material properties and biological activity of lipase within sol–gel derived bioglasses via organosilane and polymer doping. *Chem. Mater.* 2000, 12, 3695–3704.

[83] Chen, Q.; Kenausis, G. L.; Heller, A. Stability of oxidases immobilized in silica gels, *J. Am. Chem. Soc.* 1998, 120, 4582–4585.

[84] Heller, J.; Heller, A. Loss of activity or gain in stability of oxidases upon their immobilization in hydrated silica: significance of the electrostatic interactions of surface arginine residues at the entrances of the reaction channels. *J. Am. Chem. Soc.* 1998, 120, 4586–4590.

[85] Chen, X.; Jia, J.; Dong, S. Organically modified sol–gel chitosan composite based glucose biosensor. *Electroanalysis* 2003, 15, 608–612.

[86] Pandey, P. C.; Upadhyay, S.; Pathak, H. C. A new glucose sensor based on encapsulated glucose oxidase with in organically modified sol–gel glass. *Sens. Actuators* B 1999, 60, 83–89.

[87] Pandey, P. C.; Upadhyay, S.; Tiwari, I.; Tripathi, V. S. An ormosil based peroxide biosensor—a comparative study on direct electron transport from horseradish peroxidase. *Sens. Actuatuators* B 2001, 72, 224–232.

[88] Pandey, P. C.; Upadhyay, S.; Pathak, H. C. Tiwari, I. Acetylthiocholine/ acetylcholine and thiocholine/choline electrochemical biosensors/sensors based on an organically modified sol–gel glass enzyme reactor and graphite paste electrode. *Sens. Actuators* B 2000, 62, 109–116.

[89] Wang, J. Sol–gel materials for electrochemical biosensors, *Anal. Chim. Acta* 1999, 399, 21–27.

[90] Pandey, P. C.; Upadhyay S.; Shukla, N. K.; Sharma, S. Studies on the electrochemical performance of glucose oxidase modified graphite paste electrode. *Biosens. Bioelectron.* 2003, 18, 1257-1268.

[91] Pankratov, I.; Lev, O. Sol-gel derived renewable-surface biosensors. *J. Electroanal. Chem.* 1995, 393, 35-41.

[92] Hussain, F.; Birch, D. J.; Pickup, J. C. Glucose sensing based on the intrinsic fluorescence of sol-gel immobilized yeast hexokinase. *Anal. Biochem.* 2005, 339, 137-143.

[93] Pickup, J. C.; Hussain F.; Evans M. D.; Rolinski, O. J.; Birch, D. J. S. Fluorescence-based glucose sensors. *Biosens. Bioelectron.* 2005, 20, 2555-2565.

[94] Pickup, J. C.; Hussain, F.; Evans, N. D., Sachedina, N. In vivo glucose monitoring: the clinical reality and the promise. *Biosens. Bioelectron.* 2005, 20, 1897-1902.

[95] Mansur, H. S.; Lobato, Z. P.; Orefice, R.; Vasconcelos, W. L.; Machado, L. J.C. Surface functionalization of porous glass networks: effects of bovine serum albumin and porcine insulin immobilization. *Biomacromolecules*, 2000, 1, 789-797.

[96] Mansur, H. S.; Lobato, Z. P.; Orefice, R. L.; Vasconcelos, W. L.; Machado L. J. C.; Mansur, E. Adsorption/desorption behavior of bovine serum albumin and porcine insulin on chemically patterned porous gel networks. *Adsorption*, 2001, 7,105-116.

[97] Noureddini, H.; Gao, X. Characterization of sol–gel immobilized lipases. *J. Sol–Gel Sci. Technol.* 2007, 41, 31–41.

[98] Soares, C. M. F.; dos Santos, H. F.; Itako, J. E.; de Moares, F. F.; Zanin, G. M. NMR characterization of the role of silane precursors on the catalytic activity of sol–gel encapsulated lipase. *J. Non-Cryst Solids*, 2006, 352, 3469–3477.

[99] Chirico, G.; Cannone, F.; Beretta, S.; Diaspro, A., Campanini, B.; Bettati, S.; Ruotolo, R.; Mozzarelli, A. Dynamics of green fluorescent protein mutant2 in solution, on spin-coated glasses, and encapsulated in wet silica gels. *Protein Sci.* 2002, 11, 1152–1161.

[100] Zheng, L.; Brennan, J. D. Measurement of intrinsic fluorescence to probe the conformational stability and thermodynamic stability of a single tryptophan protein entrapped in a sol–gel derived glass matrix. *Analyst* 1998, 123, 1735–1744.

[101] Gottfried, D. S.; Kagan, A.; Hoffman, B. M.; Friedman, J. M. Impeded rotation of a protein in a sol–gel matrix. *J. Phys. Chem.* 1999, 103, 2803–2807.

[102] Sing, K. S. W.; Everett, D. H.; Haul, R. A. W.; Moscou, L.; Pierotti, R. A., Rouquerol, J.; Siemieniewska, T. Reporting physisorption data for gas/solid systems with special reference to the determination of surface area and porosity (Recommendations 1984). *Pure Appl. Chem.* 1985, 57, 603–619.

[103] Everett, D. H.; Powl, J. C. Adsorption in slit-like and cylindrical micropores in the Henry's law region. A model for the microporosity of carbons. *J. Chem Soc, Faraday Trans* 1 1976, 72, 619–636.

[104] El Rassy, H.; Pierre, A. C. NMR and IR spectroscopy of silica aerogels with different hydrophobic characteristics. *J. Non-Cryst. Solids* 2005, 351, 1603–1610.

[105] Engelhardt, G.; Michel, D. In: High Resolution Solid State NMR of Silicates and Zeolites.; Wiley and Sons: New York, USA, 1987.

[106] Ogura, K.; Nakaoka, K.; Nakayama, M.; Kobayashi, M.; Fuji, A. Thermogravimetry/mass spectrometry of urease-immobilized sol–gel silica and the application of such a urease-modified electrode to the potentiometric determination of urea. *Anal. Chim. Acta* 1999, 384, 219–225.

[107] Yoldas, B. E. A. Transparent porous alumina. *Bull. Am. Ceram. Soc.* 1975, 54, 286–295.

[108] Brinker, C. J.; Keefer, K. D.; Schaefer, D. W.; Assink, C. S. Sol–gel transition in simple silicate. *J. Non-Cryst. Solids* 1982, 48, 47– 64.

[109] Brinker, C. J.; Keefer, K. D.; Schaefer, D. K. W.; Assink, R. A.; Kay, B. D.; Ashley, C. S. Sol–gel transition in simple silicate II. *J. Non-Cryst. Solids* 1984, 63, 45–59.

[110] Zallen, R. The Physics of Amorphous Solids; John Wiley and Sons: New York, USA, 1983; pp 86-129 .

[111] Wright, A.C. Scientific opportunities for the study of amorphous solids using pulsed neutron sources. *J. Non-Cryst. Solids* 1985, 76, 187–210.

[112] Rodgers, L. E.; Holden, P. J.; Knott, R. B.; Finnie, K. S.; Bartlett, J. R.; Foster, J. R. Effect of sol–gel encapsulation on lipase structure and function: a small-angle neutron scattering study. *J. Sol–Gel Sci. Technol.* 2005, 33, 65–69.

[113] Edmiston, P. L.; Wambolt, C. L.; Smith, M. K.; Saavedra, S. S. Spectroscopic characterization of albumin and myoglobin entrapped in bulk sol-gel glasses. *J. Colloid Interface Sci.* 1994, 163, 395-406.

[114] Samuni, U.; Dantsker, D.; Khan, I.; Friedman, A. J.; Peterson, E.; Friedman, J.M. Spectroscopically and kinetically distinct conformational populations of sol-gel-encapsulated carbonmonoxy myoglobin. A comparison with hemoglobin. *J. Biol. Chem.* 2002, 277, 25783-25790.

[115] Juszczak, L. J.; Friedman, J. M. UV resonance Raman spectra of ligand binding intermediates of sol-gel encapsulated hemoglobin. *J. Biol. Chem.* 1999, 274, 30357–30360.

[116] West, J. M.; Kantrowitz, E. R. Trapping specific quaternary states of theallosteric enzyme aspartate transcarbamoxylase in the silica matrix sol-gels. *J. Am. Chem. Soc.* 2003, 125, 9924-9925.

[117] Shibayama, N.; Saigo, S. Fixation of the quaternary structure of human adult haemoglobin by encapsulation in transparent porous silica gel. *J. Mol. Biol.* 1995, 251, 203-209.

[118] Shibayama, N.; Saigo, S. Kinetics of the allosteric transition in hemoglobin within silicate sol-gels. *J. Am. Chem. Soc.* 1999, 121, 444-445.

[119] McIninch, J.; Kantrowitz, E. R. Use of silicate sol-gel to trap the R and T quaternary conformational states of pig kidney fructose-1, 6-bisphosphatase. *Biochem. Biophys. Acta* 2001, 1547, 320-328.
[120] Khan, I.; Shannon, C. F.; Dantsker, D.; Friedman, A. J.; Perez-Gonzalez-de-Apodaca, J.; Friedman, J. M. Sol-gel trapping of functional intermediates of hemoglobin: Geminate and bimolecular recombination studies. *Biochemistry* 2000, 39, 16099–16109.
[121] Viappiani, C.; Bettati, S.; Bruno, S.; Ronda, L.; Abbruzzetti, S.; Mozzarelli, A.; Eaton, W.A. New insights into allosteric mechanism from trapping unstable protein conformations in silica gels. *Proc. Natl. Acad. Sci. USA* 2004 101, 14414–14419.
[122] Baldini, G.; Cannone F.; Chirico, G. Pre-unfolding resonant oscillations of single green fluorescent protein molecules. *Science.* 2005, 309, 1096–100.
[123] Cannone, F.; Bologna, S.; Campanini, B.; Diaspro, A.; Bettati, S.; Mozzarelli, A.; Chirico, G. Tracking unfolding and refolding of single GFPmut2 molecules. *Biophys. J.* 2005, 89, 2033–45.
[124] Cannone, F.; Collini, M.; Chirico, G.; Baldini, G.; Bettati, S.; Campanini, B.; Mozzarelli, A. Environment effects on the oscillatory unfolding kinetics of GFP. *Eur. Biophys. J.* 2007, 36, 795–803.
[125] Bettati, S.; Mozzarelli, A. T-state hemoglobin binds oxygen noncooperatively with allosteric effects of protons, inositol hexaphosphate, and chloride. *J. Biol. Chem.* 1997, 272, 32050–32055.
[126] Bruno, S.; Bonaccio, M.; Bettati, S.; Rivetti, C.; Viappiani, C.; Abbruzzetti, S.; Mozzarelli, A. High and low oxygen affinity conformations of T-state hemoglobin. *Protein Sci.* 2001, 10, 2401–2407.
[127] Shibayama, N. Functional analysis of hemoglobin molecules locked in doubly liganded conformations. *J. Mol. Biol.* 1999, 285, 1383–1388.
[128] Khan, I.; Dantsker, D.; Samuni, U.; Friedman, A. J.; Bonaventura, C.; Manjula, B.; Acharya, S.A.; Friedman, J.M. Beta 93 modified hemoglobin: kinetic and conformational consequences. *Biochemistry.* 2001, 40, 7581–7592.
[129] Hughson, F. M.; Wright, P. E.; Baldwin. R. L. Structural characterization of a partly folded apomyoglobin intermediate. *Science.* 1990, 249,1544–1548.
[130] Hughson, F.M.; Barrick, D.; Baldwin R.L. Probing the stability of a partly folded apomyoglobin intermediate by site-directed mutagenesis. *Biochemistry.* 1991, 30, 4113–4118.
[131] Barrick, D., Baldwin. R. L. Three-state analysis of sperm whale apomyoglobin folding. *Biochemistry.* 1993, 32, 3790–3796.

[132] Jennings, P. A.; Wright, P. E. Formation of a molten globule intermediate early in the kinetic folding pathway of apomyoglobin. *Science.* 1993, 262, 892–896.
[133] Eliezer, D.; Wright. P. E. Is apomyoglobin a molten globule? Structural characterization by NMR. *J. Mol. Biol.* 1996, 263, 531–538.
[134] Shin, H. C.; Merutka, G.; Waltho, J. P.; Tennant, L. L.; Dyson, H.J.; Wright, P.E.. Peptide models of proteins folding initiation sites. 3. The G-H helical hairpin of myoglobin. *Biochemistry.* 1993, 32, 6356–6364.
[135] Sanctis, G. D.; Ascoli, F.; Brunori. M. Folding of apominimyoglobin. *Proc. Natl. Acad. Sci. USA.* 1994, 91, 11507–11511.
[136] Ballew, R. M.; Sabelko, J.; Gruebele. M. Direct observation of fast protein folding: the initial collapse of apomyoglobin. *Proc. Natl. Acad. Sci. USA.* 1996 93, 5759–5764.
[137] Sabelko, J.; Ervin, J.; Gruebele, M. Cold-denatured ensemble of apomyoglobin: implications for the early steps of folding. *J. Phys. Chem.* 1998, 102, 1806–1819.
[138] Du, Q.; Freysz, E.; Shen. Y. R.. Vibrational spectra of water molecules at quartz/water interfaces. *Phys. Rev. Lett.* 1994, 72, 238–241.
[139] Cha, P.; Krishnan, A.; Fiore, V. F.; Vogler, E. A. Interfacial energetics of protein adsorption from aqueous buffer to surfaces with varying hydrophobicity. *Langmuir.* 2008, 24, 2553-2563.
[140] Frauenfelder, H.; Parak, F.; Young, R.D. Conformational substates in proteins. *Annu. Rev. Biophys. Biophys. Chem.* 1988, 17, 451-479.
[141] Frauenfelder, H.; Nienhaus, G.U.; Johnson, J.B. Rate processes in proteins. *Physical Chemistry–Chemical Physics.* 1991, 95, 272 - 278.
[142] Frauenfelder, H.; McMahon, B. H.; Austin, R. H.; Chu, K.; Groves, J.T. The role of structure, energy landscape, dynamics, and allostery in the enzymatic function of myoglobin. *Proc. Natl. Acad. Sci. USA.* 2001, 98, 28-30.
[143] Frauenfelder, H.; McMahon, B.H. Relaxations and fluctuations in myoglobin. *Biosystems.* 2001, 62, 3-8.
[144] Klimov, D. K.; Newfield, D.; Thirumalai, D. Simulations of beta-hairpin folding confined to spherical pores using distributed computing. *Proc. Natl. Acad. Sci. USA.* 2002, 99, 8019-8024.
[145] Thirumalai, D.; Klimov, D.K.; Lorimer, G.H. Caging helps proteins fold. *Proc. Natl. Acad. Sci. USA.* 2003, 100, 11195-11197.
[146] Campanini, B.; Bologna, S.; Speroni, F. Salsi, E.; Cook, P. F.; Roderick, S. L.; Huang, B.; Bettati, S.; Mozzarelli, A. Interaction of serine acetyltransferase with O-acetylserine sulfhydrylase active site: evidence from fluorescence spectroscopy. *Protein Sci.* 2005, 14, 1125-1133.

[147] Samuni, U.; Dantsker, D.; Juszczack, L. J.; Bettati, S.; Ronda, L.; Mozzarelli, A.; Friedman, J.M. Spectroscopic and functional characterization of T-state hemoglobin conformations encapsulated in silica gels. *Biochemistry.* 2004, 43, 13674-13682.

[148] Shibayama, N.; Saigo, S. Oxygen equilibrium properties of myoglobin locked in the liganded and unliganded conformations. *J. Am. Chem. Soc.* 2003, 125, 3780-3783.

[149] Levantino, M.; Cupane, Zimanyi, L. Quaternary structure dependence of kinetic hole burning and conformational substates interconversion in hemoglobin. *Biochemistry.* 2003, 42, 4499-4505.

[150] Ronda, L.; Bruno, S. Viappiani, C.; Abbruzetti, S.; Mozzarelli, A.; Lowe, K. C.; Bettati, S. Circular dichroism spectroscopy of tertiary and quaternary conformations of human hemoglobin entrapped in wet silica gels. *Protein Sci.* 2006, 15, 1961-1967.

[151] Pioselli, B.; S. Bettati, S.; Demidkina, T. V.; Zakomirdina, L. N. Phillips, R. S.; Mozzarelli, A. Tyrosine phenol-lyase and tryptophan indole-lyase encapsulated in wet nanoporous silica gels : selective stabilization of tertiary conformations. *Protein Sci.* 2004, 13, 913-924.

[152] Pioselli, B.; Bettati, S.; Mozzarelli, A. Confinement and crowding effects on tryptophan synthase alpha2beta2 complex. *FEBS Lett.* 2005, 579, 2197-2202.

[153] Barciszewski, J.; Jurczak, J.; Porowski, S.; Specht, T.; Erdmann, V.A. The role of water structure in conformational changes of nucleic acids in ambient and high-pressure conditions. *Eur. J. Biochem.* 1999, 260, 293-307.

[154] Kauzmann, W. Some factors in the interpretation of protein denaturation. *Adv. Protein Chem.* 1959, 14, 1-63.

[155] Tanford, C. The Hydrophobic Effect: Formation of Micelles and Biological Membranes, 2nd Ed.; Wiley Interscience: New York, USA, 1980; 233pp.

[156] Pratt, L. R. Molecular theory of hydrophobic effects: She is too mean to have her name repeated. *Annu. Rev. Phys. Chem.* 2002, 53, 409-436.

[157] Sorin, E. J.; Pande. V. S. Nanotube Confinement Denatures Protein Helices. *J. Am. Chem. Soc.* 2006, 128: 6316-6317.

[158] Sorin, E. J., Rhee, Y. M.; Shirts, M. R.; Pande, V. S. The Solvation Interface is a determining Factor in Peptide Conformational Preferences. *J. Mol. Biol.* 2006, 356, 248-256.

[159] Zhou, H-X. Helix formation inside a nanotube: possible influence of backbonewater hydrogen bonding by the confining surface through modulation of water activity. *J. Chem. Phys.* 2007, 127, 245101

[160] Zhou, R. Free energy landscape of protein folding in water: explicit vs. implicit solvent. *Proteins*. 2003, 53,148–161.
[161] Zhou, R.; Huang, X.; Margulis, C. J.; Berne B. J Hydrophobic collapse in multidomain protein folding. *Science*. 2004, 305, 1605–1609.
[162] Clifford, J. S.; Legge, R. L. Use of water to evaluate hydrophobicity of organically modified xerogel enzyme supports. *Biotechnol. Bioeng*. 2005, 92, 231–237.
[163] Böhm, G.; Muhr, R.; Jaenicke, R. Quantitative analysis of protein far UV circular dichroism spectraby neural networks. *Protein Eng*. 1992, 5, 191-195.
[164] Presta, L. G.; Rose, G. D. Helix signals in proteins. *Science*. 1988, 240, 1632-1641.
[165] Mah, S. K.; Chung, I. J. Effects of dimethyldimethoxysilane on tetraethylorthosilicate sol-gel process. *J. Non-Cryst. Solids*. 1995, 183, 252.
[166] Duval, Y.; Mielczarski, J. A.; Pocrowsky, O. S.; Mielczarski, E.; Ehrhardt, J.J. Evidence of the existence of three types of species at the quartz-aqueous solution interface at pH 0-10: XPS surface group quantification and surface complexation modelling. *J. Phys. Chem. B* 2002, 106, 2937-2945.
[167] Maste, M. C. L.; Pap, E. H. W.; van Hoek, A.; Norde, W.; Visser. A. J. W. G. Spectroscopic investigation of the structure of a protein adsorbed on a hydrophobic latex. *J. Colloid Interface Sci*. 1996, 180, 632–633.
[168] Maste, M. C. L.; Norde, W.; Visser. A. J. W. G. Adsorption induced conformational changes in the serine proteinase savinase: a tryptophan fluorescence and circular dichroism study. *J. Colloid Interface Sci*. 1997, 196, 224–230.
[169] Zoungrana, T.; Findenegg, G. H.; Norde. W. Structure, stability, and activity of adsorbed enzymes. *J. Colloid Interface Sci*. 1997, 190, 437–448.
[170] Wu, H.; Fan, Y.; Sheng, J.; Sui. S.-F. Induction of changes in the secondary structure of globular proteins by a hydrophobic surface. *Eur. Biophys. J*. 1993, 22, 201–205.
[171] Baldwin, R. L. How Hofmeister ion interactions affect protein stability. *Biophys. J*. 1996, 71, 2056-2063.
[172] Collins, K. D. Charge density-dependent strength of hydration and biological structure. *Biophys. J*. 1997, 72, 65–76.
[173] Collins, K. D.; Washabaugh, M. W. The Hofmeister effect and the behavior of water at interfaces. *Q. Rev. Biophys*. 1985, 18, 323–422.
[174] Hofmeister, F. Zur Lehre Von Der Wirkung Der Salze. Zweite Mittheilung. *Arch. Exp. Pathol. Pharmakol*. 1888, 24, 247-260.

[175] Hribar, B.; Southall, N. T.; Vlacy, V.; Dill, K. A. How ions affect the structure of water. *J. Am. Chem. Soc.*, 2002, 124, 12302- 12311.
[176] Zou, Q.; Bennion, B. J.; Daggett, V.; Murphy, K. P. The molecular mechanism of stabilization of proteins by TMAO and its ability to counteract the effects of urea. *J. Am. Chem. Soc.*, 2002, 124, 1192- 1202.
[177] Batchelor, J. D.; Olteanu, A.; Tripathy, A.; Pielak, G. J. Impact of protein denaturants and stabilizers on water structure. *J. Am. Chem. Soc.* 2004, 126, 1958-1961.
[178] Mancinelli, R.; Botti, A.; Bruni, F.; Ricci, M. A.; Soper, A. K. Perturbation of water structure due to monovalent ions in solution. *Phys. Chem. Chem. Phys.*, 2007, 9, 2959 – 2967.
[179] Ellis, R. J; Minton, A. P. 2006. Protein aggregation in crowded environments. *Biol. Chem.* 387, 485–497.
[180] Hall, D.; Minton, A. P. Macromolecular crowding: qualitative and semiquantitative successes, quantitative challenges. *Biochim. Biophys. Acta.* 2003, 1649, 127–139.
[181] Minton, A. P. Influence of macromolecular crowding upon the stability and state of association of proteins: predictions and observations. *J. Pharm. Sci.* 2005, 94, 1668–1675.
[182] Minton, A. P. How can biochemical reactions within cells differ from those in test tubes? *J. Cell Sci.* 2006, 119, 2863–2869.
[183] Rivas, G; Ferrone, F.; Herzfeld, J. Life in a crowded world. *EMBO Rep.* 2004, 5, 23–27.
[184] Zhou, H-X. Loops, linkages, rings, catenanes, cages, and crowders: entropy-based strategies for stabilizing proteins. *Acc. Chem. Res.* 2004, 37, 123–130.
[185] Zhou, H-X. Protein folding in confined and crowded environments. *Arch. Biochem. Biophys.* 2008, 469, 76–82.
[186] Simionescu, N.; Simionescu, M.; Palade, G. E. Permeability of muscle capillaries to exogenous myoglobin. *J. Cell Biol.* 1973, 57, 423–52.
[187] Voet, D.; Voet, J. G.; Pratt, C.W. Fundamentals of biochemistry. John Wiley and Sons: New York, USA, 2001; p. 168.
[188] Tu, R. S.; Breedveld, V. Microrheological detection of protein unfolding. *Physical Review E.*, 2005. 72, 041914-041919.
[189] Fischer, T.; Hess, H. Materials chemistry challenges in the design of hybrid bionanodevices: supporting protein function within artificial environments. *J. Mater. Chem.* 2007, 17, 943–951.
[190] Jaeckel, C.; Salwiczek, M.; Koksch, B. Fluorine in a native protein environment—How the spatial demand and polarity of fluoroalkyl groups affect protein folding. *Angew. Chem. Int. Ed.* 2006, 45, 4198 –4203.

[191] Yoder, N. C.; Kumar, K. Fluorinated amino acids in protein design and engineering. *Chem. Soc. Rev.* 2002, 31, 335 – 341.
[192] Biffinger, J. C.; Kim, H. W.; DiMagno, S. G. The polar hydrophobicity of fluorinated compounds. *ChemBioChem.* 2004, 5, 622 – 627.
[193] Gerebtzoff, G.; Li-Blatter, X.; Fischer, H.; Frentzel, A.; Seelig, A. Halogenation of drugs enhances membrane binding and permeation. *ChemBioChem.* 2004, 5, 676 – 684.
[194] Olsen, J. A.; Banner, D.W.; Seiler, P.; Wagner, B.; Tschopp, T.; Obst-Sander, U.; Kansy, M.; Muller, K.; Diederich, F. Fluorine interactions at the thrombin active site: protein backbone fragments H-C(alpha)-C=O comprise a favorable C-F environment and interactions of C-F with electrophiles. *Chem. Bio. Chem.* 2004, 5, 666 – 675.
[195] Smart, B. E. Fluorine substituents effects (on bioactivity). *J. Fluorine Chem.* 2001, 109, 3 – 11.
[196] Dunitz, J. D. Organic fluorine: old man out. *Chem .Bio. Chem.* 2004, 5, 614 – 621.
[197] Jaeckel, C.; Koksch, B. Fluorine in peptide design and protein engineering. *Eur. J. Org. Chem.* 2005, 4483 – 4503.
[198] Leroux, F. Atropisomerism, biphenyls, and fluorine: a comparison of rotational barriers and twist angles. *Chem. Bio. Chem.* 2004, 5, 644 – 649.
[199] Mikami, K.; Itoh, Y.; Yamanaka, M. Fluorinated carbonyl and olefinic compounds: basic character and asymmetric catalytic reactions. *Chem. Rev.* 2004, 104, 1 – 16. 1995, 781 – 786.
[200] Zanda, M. Trifluoromethyl group: an effective xenobiotic function for peptide backbone modification. *New J. Chem.* 2004, 28, 1401 – 1411.
[201] Menaa, B.; Menaa, F; Avakyants, L.; Sharts, O. Protein encapsulation in nanoporous silica-based sol-gel glass: apomyoglobin secondary structure enhanced by the presence of fluoro-phosphonate groups. 2009, submitted.
[202] Gardeniers, H. J. G. E. Chemistry in nanochannel confinement. *Anal. Bioanal. Chem.* 2009, 394, 385-397.
[203] Ha, J.; Wolf, J. H.; Hillmyer, M. A.; Ward, M. D. Polymorph selectivity under nanoscopic confinement. *J. Am. Chem. Soc.* 2004, 126, 3382-3883.
[204] Matsuda, K.; Hibi, T.; Kadowaki, H.; Kataura, H.; Maniwa, Y. Water thermodynamics inside single-wall carbon nanotubes: NMR observations. *Phys. Rev. B.* 2006, 74, 073415/1-073415/4.
[205] Kresge, C. T.; Leonowicz, M. E.; Roth, W. J.; Vartuli, J.C.; Beck, J.S. Ordered mesoporous molecular sieves synthesized by a liquid-crystal template mechanism. *Nature.* 1992, 359, 710-712.

[206] Thomas, J. M. The chemistry of crystalline sponges. *Nature.* 1994, 368, 289-290.
[207] Hunger, M.; Horvath, T. A new MAS NMR probe for in situ investigations of hydrocarbon conversion on solid catalysts under continuous-flow conditions. *J. Chem. Soc., Chem. Commun.* 1995, 1423-1424.
[208] Haw, J. F.; Nicolas, J. B.; Xu, T.; Beck, L. W.; Ferguson, D. B. Physical organic chemistry of solid acids: lessons from in situ NMR and theoretical chemistry. *Acc. Chem. Res.* 1996, 29, 259-267.
[209] Menaa, F.; Boucharaba, A.; Menaa, B.; Guimarães, C.A.; Avakyants, L.; Sharts, O. Fluoro-Raman spectroscopy as a new analytical tool to detect ex-vivo oncoproteins in tumor cell lines, submitted.
[210] Dave, B. C.; Miller, J. M.; Dunn, B.; Valentine, J.S.; and Zink, J.I. Encapsulation of proteins in bulk and thin film sol gel matrices. *J. Sol-Gel Sci. Technol.* 1997, 8, 629-634.
[211] Peterson, E. S.; Shinder, R.; Khan, I.; Juczszak, L.; Wang, J.; Manjula, B.; Acharya, S.A.; Bonaventura, C.; Friedman, J.M. Domain-specific effector interactions within the central cavity of human adult hemoglobin in solution and in porous sol gel matrices: evidence for long-range communication pathways. *Biochemistry.* 2004, 43, 4832-4843.
[212] Kato, K.; Gong, Y.; Saito, T.; Yokogawa, Y. Preparation and catalytic performance of lipases encapsulated in sol-gel materials. *Biosci. Biotechnol. Biochem.* 2002, 66, 221-223.

INDEX

A

accessibility, 35
acetone, 12
acetylcholine, 57
acetylcholinesterase, 10
acid, 5, 6, 17, 40, 41, 42, 47, 54
acidity, 39
active oxygen, 26
active site, 61, 65
adaptation, 44
additives, 10
adjustment, 6
adsorption, 11, 12, 13, 29, 34, 43, 51, 61
aerogels, 7, 9, 11, 56, 58
AFM, 12
aggregation, 36, 52, 64
aging, 7, 8, 56
albumin, 59
alcohol, 6, 10, 51, 55
algorithm, 24
ALS, 15
alters, 22
aluminum, 12
amino acids, 18, 39, 65
anisotropy, 16
ANS, 15
anticancer drug, 10, 52
arginine, 57
aspartate, 16, 59
atoms, 28, 29, 32
ATP, 56

B

bacteria, 10
barriers, 20, 65
basicity, 39
behavior, 30, 58, 63
binding, 9, 15, 16, 40, 59, 65
bioactive materials, 2
biocatalysts, vii, ix, 40, 48, 52, 53, 56
biochemistry, 64
biocompatibility, 10, 47
biological activity, vii, ix, 3, 18, 22, 25, 35, 57
biological systems, 6, 15, 20
biomaterials, x, 1, 37, 47, 56
bionanodevices, vii, ix, 47, 64
biosensors, 2, 10, 40, 48, 51, 55, 57
biotechnology, 6, 39
blood, 39, 44
bonding, 21, 62
bonds, 39, 45
brain, 39
buffer, 6, 7, 8, 16, 23, 24, 42, 61
burning, 62

C

calorimetry, 20
cancer, 10
candida, 53
capillary, 11, 44
carbon, 21, 39, 44, 54, 65

carbon nanotubes, 21, 54, 65
carrier, 52
catalyst, 5, 47
catalytic activity, 2, 9, 16, 35, 48, 58
cation, 9
cell, 7, 20, 32, 52, 66
ceramic, 52, 54
channels, 57
chemical interaction, 55
chemical properties, 6, 11
chemotherapy, 44
circularly polarized light, 3, 18
classes, 6
classification, 12, 35
CO_2, 8, 12
communication, 66
community, x
compatibility, 6
complement, 29
composition, 2, 5, 7, 17, 24, 25, 29, 33
compounds, 6, 44, 52, 54, 65
comprehension, 48
concentration, 6, 18, 30, 31, 44
condensation, 15, 26, 28
configuration, 48
confinement, 7, 16, 21, 33, 37, 52, 56, 65
connectivity, 21
control, 1, 2, 9, 10, 23, 49
conversion, 11, 18, 53, 66
correlation, 34, 35
crack, 56
crystal structure, 34
crystalline, 54, 56, 66
cytochrome, 9
cytoplasm, ix, 1, 32, 33
cytoskeleton, 33

D

DCA, 10
deconvolution, 24
dehydration, 10, 37
delivery, ix
denaturation, 1, 9, 36, 37, 47, 62
density, 63

desorption, 13, 34, 58
detection, 44, 45, 56, 64
diabetes, 10
diffusion, 7, 39, 48
dispersion, 10, 54
distillation, 6, 47
distributed computing, 61
distribution, 12, 13, 16
DNA, 10
domain structure, 14
doping, 57
drug delivery, vii, 1, 10, 43
drugs, 39, 65
drying, 7, 8, 10, 11, 12, 33, 34, 56
DSC, 20
dyes, 6

E

electrochemistry, 54
electron, 57
encapsulation, ix, 1, 2, 6, 8, 9, 11, 12, 13, 15, 16, 17, 20, 26, 29, 34, 40, 47, 51, 53, 55, 59, 65
endothermic, 21
energy, 19, 20, 28, 61, 63
entrapment, 6, 9, 10, 15, 16, 53, 56
entropy, 21, 29, 64
environment, ix, x, 1, 2, 9, 10, 11, 16, 17, 20, 21, 22, 25, 26, 27, 33, 34, 40, 43, 49, 51, 55, 64, 65
environmental change, 49
enzymatic activity, 9
enzymes, 2, 6, 9, 10, 11, 17, 20, 22, 30, 35, 48, 53, 55, 63
equilibrium, 16, 28, 62
ethanol, 5
ethylene glycol, 10
eukaryotic cell, 32
evaporation, 6, 12
evolution, 43, 56
experimental condition, 40, 48
extraction, 8

Index

F

films, 54
fluctuations, 61
fluid, 12
fluorescence, 1, 3, 7, 15, 16, 35, 54, 57, 58, 61, 63
fluorine, 2, 3, 7, 23, 26, 39, 40, 41, 42, 44, 48, 49, 65
fluorine atoms, 40
fractal dimension, 15
free energy, 19, 28, 31, 48, 49
fructose, 60
FTIR, 15
functionalization, 22, 23, 28, 58

G

gel, ix, 1, 2, 3, 5, 6, 7, 8, 9, 10, 11, 15, 16, 17, 19, 20, 21, 22, 32, 33, 47, 51, 52, 53, 54, 55, 56, 57, 58, 59, 60, 66
gelation, 6, 14
generation, 52
glasses, vii, ix, x, 1, 2, 3, 6, 7, 11, 13, 14, 16, 17, 19, 20, 21, 22, 23, 24, 25, 26, 28, 29, 30, 31, 32, 33, 34, 35, 37, 40, 41, 44, 47, 48, 49, 51, 53, 54, 55, 58, 59
glucose, 10, 57, 58
glucose oxidase, 10, 57
graphite, 57
groups, vii, ix, 1, 2, 3, 5, 7, 9, 14, 17, 19, 20, 21, 24, 25, 26, 27, 28, 29, 31, 32, 34, 39, 40, 41, 42, 44, 48, 49, 64, 65
growth, 15
growth mechanism, 15

H

halogen, 54
heat, 36
heating, 36
height, 20
helical conformation, 26
helicity, 18, 21, 23, 24, 26, 28, 29, 31, 33, 34, 37, 41, 42, 48, 49
heme, 2, 17, 26, 48, 51
hemoglobin, 16, 59, 60, 62, 66
host, vii, ix, 1, 2, 7, 8, 9, 11, 12, 14, 16, 17, 18, 20, 22, 33, 34, 35, 40, 43, 44, 45, 47, 48, 49
hybrid, 1, 5, 6, 7, 8, 14, 40, 47, 54, 55, 64
hydrogen, 3, 19, 21, 28, 31, 39, 43, 45, 49, 62
hydrogen bonds, 3, 28, 31, 39, 43, 45, 49
hydrolysis, 1, 5, 6, 8, 10, 47, 48, 53
hydrophilicity, 9, 10, 31
hydrophobicity, vii, ix, 2, 3, 9, 10, 14, 20, 22, 23, 24, 29, 31, 32, 35, 39, 41, 44, 47, 48, 49, 61, 63, 65
hypothesis, 11, 22, 28, 35
hysteresis, 12, 13, 34

I

identification, 14
immobilization, 6, 10, 47, 57, 58
immobilized enzymes, 10, 15
implementation, 6
in vivo, 10
industry, 37
infrared spectroscopy, 22
initiation, 61
inositol, 60
insight, 3, 17
instability, 16
insulin, 58
interaction, 12
interactions, 3, 14, 19, 20, 27, 28, 29, 30, 39, 45, 48, 53, 55, 57, 63, 65, 66
interface, 21, 25, 26, 28, 29, 31, 35, 63
interference, 6
ions, 12, 20, 29, 30, 31, 32, 40, 47, 64
IR spectroscopy, 58
isotherms, 12, 13, 34

K

K^+, 30
kidney, 60

kinetics, 16, 17, 52, 60

L

leaching, 9, 36
ligand, 15, 59
light scattering, 15, 35
linkage, 2, 15, 25, 26, 28, 48
lipases, 2, 11, 22, 35, 48, 52, 53, 56, 58, 66
liquid phase, 20
luciferase, 56

M

macromolecules, 20, 32, 33
management, 10
MAS, 14, 17, 22, 26, 43, 44, 66
mass spectrometry, 59
materials science, 1, 44
matrix, vii, ix, 1, 2, 5, 6, 8, 9, 11, 12, 14, 16, 17, 18, 20, 22, 27, 28, 30, 33, 34, 40, 44, 45, 47, 48, 51, 55, 56, 58, 59
measurement, 18
media, 7, 52
membranes, 39
mesoporous materials, 13
methanol, 5
methyl groups, 2, 25, 28, 33
microenvironments, 11
microscopy, 11, 12
microstructure, vii, ix, 12, 14
mixing, 6
models, 20, 33, 61
molar ratios, 7
molecular weight, 18
molecules, 12, 16, 20, 22, 27, 29, 30, 32, 43, 44, 47, 54, 60, 61
molybdenum, 56
monomers, 5
morphology, 12, 14
mutagenesis, 60
mutant, 16
myoglobin, 9, 15, 16, 17, 48, 59, 61, 62, 64

N

nanomaterials, 5
nanoparticles, 10, 52
nanotube, 62
natural polymers, 10
Netherlands, 55
network, vii, ix, 5, 7, 9, 12, 14, 15, 19, 23, 25, 26, 28, 29, 33, 37, 47, 48, 53
neural networks, 63
NMR, 3, 11, 14, 15, 17, 22, 26, 35, 43, 49, 54, 56, 58, 59, 61, 65, 66
nucleic acid, 44, 62

O

observations, 54, 64, 65
oil, 53
oils, 11
organic solvents, 47, 53
oxygen, 15, 16, 28, 29, 32, 60

P

parameter, 6, 7, 28
particles, 6, 13, 14, 27, 52
partition, 28
pathways, 66
peptides, 21, 39
permeation, 65
permit, 1
peroxide, 57
PET, 7, 8
pH, 6, 7, 17, 23, 24, 26, 30, 36, 41, 63
phenol, 62
photoluminescence, 54
photonics, 6
physical properties, vii, ix, 2, 9, 17
physics, 52
platelets, 12, 35
polar groups, 21
polarity, 9, 40, 64
polarizability, 39
polycondensation, 1, 5, 7, 8, 47

Index

polycondensation process, 7
polymer, 10, 15, 33, 53, 57
polypeptide, 11, 15
porosity, vii, ix, 2, 3, 17, 20, 47, 58
potassium, 23, 24, 26, 31, 36, 41
precipitation, 36
pressure, 56, 62
probe, 2, 3, 7, 22, 58, 66
protein conformations, 2, 60
protein engineering, 65
protein folding, vii, ix, 1, 2, 3, 11, 12, 16, 17, 19, 20, 21, 28, 29, 31, 32, 33, 40, 43, 44, 47, 48, 52, 53, 61, 63, 64
proteinase, 63
protein-protein interactions, 1
proteins, ix, 1, 2, 3, 6, 9, 11, 16, 19, 20, 21, 22, 29, 30, 32, 33, 35, 37, 39, 40, 43, 44, 47, 48, 49, 51, 52, 54, 55, 56, 61, 63, 64, 66
protocol, 5, 17
protons, 29, 32, 60

Q

quartz, 7, 29, 44, 61, 63

R

Raman spectra, 15, 59
Raman spectroscopy, 3, 44, 66
range, 6, 9, 12, 39, 43, 66
reactants, 35
reagents, 24
real time, 15
reality, 58
recombination, 60
rehydration, 10
relationship, 24, 25
residues, 18, 57
resolution, 44
retention, 9, 48
reusability, 10
rings, 64
RNA, 10
room temperature, 1, 47

S

safety, 44
salt, 29, 30, 31
salts, 20, 32
sample, 31, 36, 44
scanning electron microscopy, 17
scattering, 15, 35, 59
seed, 14
selectivity, 55, 65
sensing, 51, 56, 57
sensitivity, 44, 47
sensors, vii, ix, 1, 7, 57
separation, 21
serine, 61, 63
serum, 9, 58
serum albumin, 9, 58
shape, 12, 35
sign, 18
signals, 40, 45, 63
silane, 7, 25, 48, 58
silicon, 2, 14, 20, 24, 26, 28, 33, 37, 55
SiO_2, 19
SiO_2 surface, 19
software, 24
sol-gel, ix, x, 1, 2, 3, 5, 6, 7, 8, 9, 10, 14, 15, 16, 17, 19, 20, 25, 27, 28, 29, 30, 33, 35, 40, 44, 45, 47, 48, 51, 53, 54, 55, 56, 57, 59, 60, 63, 65, 66
solid state, 2, 11, 44
solubility, 10
solvents, 8, 12
soybean, 53
species, 5, 44, 54, 63
specific surface, 34, 35
specificity, 44
spectroscopy, x, 1, 3, 7, 15, 16, 17, 22, 25, 35, 44, 61, 62
spectrum, 15, 23, 26, 41, 42, 44
sperm, 53, 60
spin, 58
stability, ix, 2, 6, 9, 10, 17, 20, 25, 33, 35, 37, 39, 41, 53, 57, 58, 60, 63, 64
stabilization, ,30, 62, 64
stages, 44

stock, 18
storage, 9
strategies, 64
strength, 14, 19, 22, 28, 48, 63
stress, 37
substitution, 25, 39
substrates, 16, 22
sugar, 9, 56, 57
surface area, 9, 11, 12, 15, 17, 20, 34, 39, 58
surface modification, x, 1, 6, 20, 22, 48
surface properties, 11, 18, 48
surface tension, 12, 56
synthesis, 6, 55, 56

T

temperature, 6, 8, 11, 35, 36
TEOS, 5, 9, 10
therapy, 52
thermal evaporation, 7
thermal stability, vii, ix, 20, 35, 36, 42, 48
thermodynamics, 3, 16, 19, 31, 33, 54, 65
thermogravimetry, 15
thrombin, 65
time resolution, 44
transition, 16, 20, 36, 59
transparency, 7, 9, 48
transport, 16, 57
tryptophan, 35, 58, 62, 63

tumor, 66
twist, 65
tyrosine, 35, 57, 62

U

uniform, 7
urea, 59, 64
UV, 15, 18, 23, 25, 40, 41, 59, 63

V

values, 6
vanadium, 56

W

workers, 9, 20, 35

X

XPS, 63

Y

yeast, 57